Introduction to Matrix Theory

Arindama Singh

Introduction to Matrix Theory

Ane Books
Pvt. Ltd.

Arindama Singh
Department of Mathematics
Indian Institute of Technology Madras
Chennai, India

ISBN 978-3-030-80483-1   ISBN 978-3-030-80481-7   (eBook)
https://doi.org/10.1007/978-3-030-80481-7

Jointly published with Ane Books Pvt. Ltd.
In addition to this printed edition, there is a local printed edition of this work available via Ane Books in
South Asia (India, Pakistan, Sri Lanka, Bangladesh, Nepal and Bhutan) and Africa (all countries in the
African subcontinent).
ISBN of the Co-Publisher's edition: 978-93-86761-20-0

This Springer imprint is published by the registered company Springer Nature Switzerland AG
The registered company address is: Gewerbestrasse 11, 6330 Cham, Switzerland

Preface

Practising scientists and engineers feel that calculus and matrix theory form the minimum mathematical requirement for their future work. Though it is recommended to spread matrix theory or linear algebra over two semesters in an early stage, the typical engineering curriculum allocates only one semester for it. In addition, I found that science and engineering students are at a loss in appreciating the abstract methods of linear algebra in the first year of their undergraduate programme. This resulted in a curriculum that includes a thorough study of system of linear equations via Gaussian and/or Gauss–Jordan elimination comprising roughly one month in the first or second semester. It needs a follow-up of one-semester work in matrix theory ending in canonical forms, factorizations of matrices, and matrix norms.

Initially, we followed the books such as Leon [10], Lewis [11], and Strang [14] as possible texts, referring occasionally to papers and other books. None of these could be used as a textbook on its own for our purpose. The requirement was a single text containing development of notions, one leading to the next, and without any distraction towards applications. It resulted in creation of our own material. The students wished to see the material in a book form so that they might keep it on their lap instead of reading it off the laptop screens. Of course, I had to put some extra effort in bringing it to this form; the effort is not much compared to the enjoyment in learning.

The approach is straightforward. Starting from the simple but intricate problems that a system of linear equations presents, it introduces matrices and operations on them. The elementary row operations comprise the basic tools in working with most of the concepts. Though the vector space terminology is not required to study matrices, an exposure to the notions is certainly helpful for an engineer's future research. Keeping this in view, the vector space terminology is introduced in a restricted environment of subspaces of finite-dimensional real or complex spaces. It is felt that this direct approach will meet the needs of scientists and engineers. Also, it will form a basis for abstract function spaces, which one may study or use later.

Starting from simple operations on matrices, this elementary treatment of matrix theory characterizes equivalence and similarity of matrices. The other tool of Gram–Schmidt orthogonalization has been discussed leading to best approximations and least squares solution of linear systems. On the go, we discuss matrix factorizations such as rank factorization, QR-factorization, Schur triangularization, diagonalization, Jordan form, singular value decomposition, and polar decomposition. It includes norms on matrices as a means to deal with iterative solutions of linear systems and exponential of a matrix. Keeping the modest goal of an introductory textbook on matrix theory, which may be covered in a semester, these topics are dealt with in a lively manner.

Though the earlier drafts were intended for use by science and engineering students, many mathematics students used those as supplementary text for learning linear algebra. This book will certainly fulfil that need.

Each section of the book has exercises to reinforce the concepts; problems have been added at the end of each chapter for the curious student. Most of these problems are theoretical in nature, and they do not fit into the running text linearly. Exercises and problems form an integral part of the book. Working them out may require some help from the teacher. It is hoped that the teachers and the students of matrix theory will enjoy the text the same way I and my students did.

Most engineering colleges in India allocate only one semester for linear algebra or matrix theory. In such a case, the first two chapters of the book can be covered in a rapid pace with proper attention to elementary row operations. If time does not permit, the last chapter on matrix norms may be omitted or covered in numerical analysis under the veil of iterative solutions of linear systems.

I acknowledge the pains taken by my students in pointing out typographical errors. Their difficulties in grasping the notions have contributed a lot towards the contents and this particular sequencing of topics. I cheerfully thank my colleagues A. V. Jayanthan and R. Balaji for using the earlier drafts for teaching linear algebra to undergraduate engineering and science students at IIT Madras. They pointed out many improvements, which I cannot pinpoint now. Though the idea of completing this work originated five years back, time did not permit it. IIT Madras granted me sabbatical to write the second edition of may earlier book on *Logics for Computer Science*. After sending a draft of that to the publisher, I could devote the stop-gap for completing this work. I hereby record my thanks to the administrative authorities of IIT Madras.

It will be foolish on my part to claim perfection. If you are using the book, then you should be able to point out improvements. I welcome you to write to me at asingh@iitm.ac.in.

Chennai, India

Arindama Singh

Contents

About the Author

Dr. Arindama Singh is a professor in the Department of Mathematics, Indian Institute of Technology (IIT) Madras, India. He received his Ph.D. degree from the IIT Kanpur, India, in 1990. His research interests include knowledge compilation, singular perturbation, mathematical learning theory, image processing, and numerical linear algebra. He has published six books, over 60 papers in journals and conferences of international repute. He has guided five Ph.D. students and is a life member of many academic bodies, including the Indian Society for Industrial and Applied Mathematics, Indian Society of Technical Education, Ramanujan Mathematical Society, Indian Mathematical Society, and The Association of Mathematics Teachers of India.

Chapter 1
Matrix Operations

1.1 Examples of Linear Equations

Linear equations are everywhere, starting from mental arithmetic problems to advanced defence applications. We start with an example. Consider the system of linear equations

$$x_1 + x_2 = 3$$
$$x_1 - x_2 = 1$$

Subtracting the first from the second, we get $-2x_2 = -2$. It implies $x_2 = 1$. That is, the original system is replaced with the following:

$$x_1 + x_2 = 3$$
$$x_2 = 1$$

Substituting $x_2 = 1$ in the first equation of the new system, we get $x_1 = 2$. We verify that $x_1 = 2$, $x_2 = 1$ satisfy the equations. Hence, the system of equations has this unique solution.

To see it geometrically, let x_1 represent points on x-axis, and let x_2 represent points on the y-axis. Then, the first equation represents a straight line that passes through the point $(3, 0)$ and has slope -1. Similarly, the second equation represents a straight line passing through the point $(1, 0)$ and having slope 1. They intersect at the point $(2, 1)$.

What about the following linear system?

$$x_1 + x_2 = 3$$
$$x_1 - x_2 = 1$$
$$2x_1 - x_2 = 3$$

© The Author(s), under exclusive license to Springer Nature Switzerland AG 2021
A. Singh, *Introduction to Matrix Theory*,
https://doi.org/10.1007/978-3-030-80481-7_1

The first two equations have a unique solution, and that satisfies the third. Hence, this system also has a unique solution $x_1 = 2$, $x_2 = 1$. Geometrically, the third equation represents the straight line that passes through $(0, -3)$ and has slope 2. The intersection of all the three lines is the same point $(2, 1)$. So, the extra equation does not put any constraint on the solutions that we obtained earlier.

But what about our systematic solution method? We aim at eliminating the first unknown from all but the first equation. We replace the second equation with the one obtained by second minus the first. We also replace the third by third minus twice the first. It results in

$$x_1 + x_2 = 3$$
$$-x_2 = -1$$
$$-3x_2 = 3$$

Notice that the second and the third equations *coincide*, hence the conclusion. We give another twist. Consider the system

$$x_1 + x_2 = 3$$
$$x_1 - x_2 = 1$$
$$2x_1 + x_2 = 3$$

The first two equations again have the solution $x_1 = 2$, $x_2 = 1$. But this time, the third is not satisfied by these values of the unknowns. So, the system has no solution.

Geometrically, the first two lines have a point of intersection $(2, 1)$; the second and the third have the intersection point as $(4/3, 1/3)$; and the third and the first have the intersection point as $(0, 3)$. They form a triangle. There is no point common to all the three lines. Also, by using our elimination method, we obtain the equations as:

$$x_1 + x_2 = 3$$
$$-x_2 = -1$$
$$-x_2 = -3$$

The last two equations are not *consistent*. So, the original system has no solution.

Finally, instead of adding another equation, we drop one. Consider the linear equation

$$x_1 + x_2 = 3$$

The old solution $x_1 = 2$, $x_2 = 1$ is still a solution of this system. But there are other solutions. For instance, $x_1 = 1$, $x_2 = 2$ is a solution. Moreover, since $x_1 = 3 - x_2$, by assigning x_2 any real number, we get a corresponding value for x_1, which together give a solution. Thus, it has infinitely many solutions.

Geometrically, any point on the straight line represented by the equation is a solution of the system. Notice that the same conclusion holds if we have more equations, which are *multiples of* the only given equation. For example,

$$x_1 + x_2 = 3$$
$$2x_1 + 2x_2 = 6$$
$$3x_1 + 3x_2 = 9$$

We see that the number of equations really does not matter, but the number of *independent* equations does matter. Of course, the notion of independent equations is not yet precise; we have some working ideas only.

It is not also very clear when does a system of equations have a solution, a unique solution, infinitely many solutions, or even no solutions. And why not a system of equations has more than one but finitely many solutions? How do we use our elimination method for obtaining infinite number of solutions?

To answer these questions, we will introduce matrices. Matrices will help us in representing the problem in a compact way and will lead to a definitive answer. We will also study the eigenvalue problem for matrices which come up often in applications. These concerns will allow us to represent matrices in elegant forms.

Exercises for Sect. 1.1

1. For each of the following system of linear equations, find the number of solutions geometrically:

   (a) $x_1 + 2x_2 = 4$, $-2x_1 - 4x_2 = 4$
   (b) $-x_1 + 2x_2 = 3$, $2x_1 - 4x_2 = -6$
   (c) $x_1 + 2x_2 = 1$, $x_1 - 2x_2 = 1$, $-x_1 + 6x_2 = 3$

2. Show that the system of linear equations $a_1x_1 + x_2 = b_1, a_2x_1 + x_2 = b_2$ has a unique solution if $a_1 \neq a_2$. Is the converse true?

1.2 Basic Matrix Operations

In the last section, we have solved a linear system by transforming it to equivalent systems. Our method of solution may be seen schematically as follows:

$$
\begin{array}{ccc}
\begin{matrix} x_1 + x_2 = 3 \\ x_1 - x_2 = 1 \end{matrix} & \Rightarrow &
\begin{matrix} x_1 + x_2 = 3 \\ x_2 = 1 \end{matrix} & \Rightarrow &
\begin{matrix} x_1 = 2 \\ x_2 = 1 \end{matrix}
\end{array}
$$

We can minimize writing by ignoring the unknowns and transform only the numbers in the following way:

$$\begin{array}{ccc} 1 & 1 & 3 \\ 1 & -1 & 1 \end{array} \quad \Rightarrow \quad \begin{array}{ccc} 1 & 1 & 3 \\ 0 & 1 & 1 \end{array} \quad \Rightarrow \quad \begin{array}{ccc} 1 & 0 & 2 \\ 0 & 1 & 1 \end{array}$$

To be able to operate with such array of numbers and talk about them, we require some terminology. First, some notation:

Notation 1.1 $\mathbb{N}$ denotes the set of all natural numbers $1, 2, 3, \ldots$
$\mathbb{R}$ denotes the set of all real numbers.
$\mathbb{C}$ denotes the set of all complex numbers.

We will write $\mathbb{F}$ for either $\mathbb{R}$ or $\mathbb{C}$. The numbers in $\mathbb{F}$ will also be referred to as **scalars**. A rectangular array of scalars is called a **matrix**. We write a matrix with a pair of surrounding square brackets as in the following.

$$\begin{bmatrix} a_{11} & \cdots & a_{1n} \\ \vdots & & \vdots \\ a_{m1} & \cdots & a_{mn} \end{bmatrix}$$

Here, a_{ij} are scalars.

We give names to matrices. If A is (equal to) the above matrix, then we say that A has m number of **rows** and n number of **columns**; and we say that A is an $m \times n$ matrix. The scalar that is common to the ith row and jth column of A is a_{ij}. With respect to the matrix A, we say that a_{ij} is its (i, j)th **entry**. The entries of a matrix are scalars. We also write the above matrix as

$$A = [a_{ij}], \quad a_{ij} \in \mathbb{F} \quad \text{for } i = 1, \ldots, m, \ j = 1, \ldots, n.$$

Thus, the scalar a_{ij} is the (i, j)th entry of the matrix $[a_{ij}]$. Here, i is called the *row index* and j is called the **column index** of the entry a_{ij}.

The set of all $m \times n$ matrices with entries from $\mathbb{F}$ will be denoted by $\mathbb{F}^{m \times n}$.

A **row vector** of size n is a matrix in $\mathbb{F}^{1 \times n}$. Similarly, a **column vector** of size n is a matrix in $\mathbb{F}^{n \times 1}$. The vectors in $\mathbb{F}^{1 \times n}$ (row vectors) will be written as (with or without commas)

$$[a_1, \ldots, a_n] \quad \text{or as} \quad [a_1 \ \cdots \ a_n]$$

for scalars $a_1, \ldots, a_n$. The vectors in $\mathbb{F}^{n \times 1}$ are written as

$$\begin{bmatrix} b_1 \\ \vdots \\ b_n \end{bmatrix} \quad \text{or as} \quad [b_1, \ldots, b_n]^t \quad \text{or as} \quad [b_1 \ \cdots \ b_n]^t$$

for scalars $b_1, \ldots, b_n$. The second way of writing is the *transpose notation*; it saves vertical space. Also, if a column vector v is equal to u^t for a row vector u, then we write the row vector u as v^t. Thus, we accept $(u^t)^t = u$ and $(v^t)^t = v$ as a way of writing.

We will write both $\mathbb{F}^{1 \times n}$ and $\mathbb{F}^{n \times 1}$ as $\mathbb{F}^n$. Especially when a result is applicable to both row vectors and column vectors, this notation will become handy. Also, we may write a typical vector in $\mathbb{F}^n$ as

$$(a_1, \ldots, a_n).$$

When $\mathbb{F}^n$ is $\mathbb{F}^{1 \times n}$, you should read $(a_1, \ldots, a_n)$ as $[a_1, \ldots, a_n]$, a row vector, and when $\mathbb{F}^n$ is $\mathbb{F}^{n \times 1}$, you should read $(a_1, \ldots, a_n)$ as $[a_1, \ldots, a_n]^t$, a column vector.

The i**th row** of a matrix $A = [a_{ij}] \in \mathbb{F}^{m \times n}$ is the row vector

$$[a_{i1}, \ldots a_{in}].$$

We also say that the **row index** of this row is i. Similarly, the j**th column** of A is the column vector

$$[a_{1j}, \ldots a_{mj}]^t.$$

And, its **column index** is j.

Any matrix in $\mathbb{F}^{m \times n}$ is said to have its **size** as $m \times n$. If $m = n$, the rectangular array becomes a square array with n rows and n columns; and the matrix is called a **square matrix** of **order** n.

Two matrices of the same size are considered **equal** when their corresponding entries coincide; i.e. if $A = [a_{ij}]$ and $B = [b_{ij}]$ are in $\mathbb{F}^{m \times n}$, then

$$A = B \text{ iff } a_{ij} = b_{ij} \text{ for } 1 \le i \le m, 1 \le j \le n.$$

Matrices of different sizes are unequal.

The **zero matrix** is a matrix, each entry of which is 0. We write 0 for all zero matrices of all sizes. The size is to be understood from the context.

Let $A = [a_{ij}] \in \mathbb{F}^{n \times n}$ be a square matrix of order n. The entries a_{ii} are called the **diagonal entries** of A. The **diagonal** of A consists of all diagonal entries; the first entry on the diagonal is a_{11}, and the last diagonal entry is a_{nn}. The entries of A, which are not on the diagonal, are called the **off-diagonal entries** of A; they are a_{ij} for $i \ne j$. In the following matrix, the diagonal is shown in bold:

$$\begin{bmatrix} \mathbf{1} & 2 & 3 \\ 2 & \mathbf{3} & 4 \\ 3 & 4 & \mathbf{0} \end{bmatrix}.$$

Here, 1 is the first diagonal entry, 3 is the second diagonal entry, and 5 is the third and the last diagonal entry.

The **super-diagonal** of a matrix consists of entries above the diagonal. That is, the entries $a_{i,i+1}$ comprise the super-diagonal of an $n \times n$ matrix $A = [a_{ij}]$. Of course, i varies from 1 to $n - 1$ here. In the following matrix, the super-diagonal is shown in bold:

$$\begin{bmatrix} 1 & 2 & 3 \\ 2 & 3 & 4 \\ 3 & 4 & 0 \end{bmatrix}.$$

If all off-diagonal entries of A are 0, then A is said to be a **diagonal matrix**. Only a square matrix can be a diagonal matrix. There is a way to generalize this notion to any matrix, but we do not require it. Notice that all diagonal entries in a diagonal matrix need not be nonzero. For example, the zero matrix of order n is a diagonal matrix. We also write a diagonal matrix with diagonal entries $d_1, \ldots, d_n$ as

$$\mathrm{diag}(d_1, \ldots, d_n).$$

The following is a diagonal matrix. We follow the convention of not showing the non-diagonal entries in a diagonal matrix, which are 0.

$$\mathrm{diag}(1, 3, 0) = \begin{bmatrix} 1 & & \\ & 3 & \\ & & 0 \end{bmatrix} = \begin{bmatrix} 1 & 0 & 0 \\ 0 & 3 & 0 \\ 0 & 0 & 0 \end{bmatrix}.$$

The **identity matrix** is a diagonal matrix with each diagonal entry as 1. We write an identity matrix of order m as I_m. Sometimes, we omit the subscript m if it is understood from the context.

$$I = I_m = \mathrm{diag}(1, \ldots, 1).$$

We write e_i for a column vector whose ith component is 1 and all other components 0. The jth component of e_i is δ_{ij}. Here,

$$\delta_{ij} = \begin{cases} 1 & \text{if } i = j \\ 0 & \text{if } i \neq j \end{cases}$$

is Kronecker's delta. The size of a column vector e_i is to be understood from the context. Notice that the identity matrix $I = [\delta_{ij}]$.

There are then n distinct column vectors $e_1, \ldots, e_n$ in $\mathbb{F}^{n \times 1}$. These are referred to as the **standard basis vectors** for reasons you will see later. We also say that e_i is the ith standard basis vector. These are the columns of the identity matrix of order n, in that order; that is, e_i is the ith column of I. Then, e_i^t is the ith row of I. Thus,

$$I = [\delta_{ij}] = \mathrm{diag}(1, \ldots, 1) = [e_1 \ \cdots \ e_n] = \begin{bmatrix} e_1^t \\ \vdots \\ e_n^t \end{bmatrix}.$$

A **scalar matrix** is a diagonal matrix with equal diagonal entries. For instance, the following is a scalar matrix:

$$\begin{bmatrix} 3 & & & \\ & 3 & & \\ & & 3 & \\ & & & 3 \end{bmatrix}.$$

It is also written as $\operatorname{diag}(3, 3, 3, 3)$.

A matrix $A \in \mathbb{F}^{m \times n}$ is said to be **upper triangular** iff all entries below the diagonal are zero. That is, $A = [a_{ij}]$ is upper triangular when $a_{ij} = 0$ for $i > j$. In writing such a matrix, we simply do not show the zero entries below the diagonal.

Similarly, a matrix is called **lower triangular** iff all its entries above the diagonal are zero.

Both upper triangular and lower triangular matrices are referred to as **triangular** matrices. In the following, L is a lower triangular matrix, and U is an upper triangular matrix, each of order 3:

$$L = \begin{bmatrix} 1 & & \\ 2 & 3 & \\ 3 & 4 & 5 \end{bmatrix}, \quad U = \begin{bmatrix} 1 & 2 & 3 \\ & 3 & 4 \\ & & 5 \end{bmatrix}.$$

A diagonal matrix is both upper triangular and lower triangular.

Sum of two matrices of the same size is a matrix whose entries are obtained by adding the corresponding entries in the given two matrices. If $A = [a_{ij}] \in \mathbb{F}^{m \times n}$ and $B = [b_{ij}] \in \mathbb{F}^{m \times n}$, then $A + B = [c_{ij}] \in \mathbb{F}^{m \times n}$ with

$$c_{ij} = a_{ij} + b_{ij} \quad \text{for } 1 \le i \le m, 1 \le j \le n.$$

We write the same thing as $[a_{ij}] + [b_{ij}] = [a_{ij} + b_{ij}]$. For example,

$$\begin{bmatrix} 1 & 2 & 3 \\ 2 & 3 & 1 \end{bmatrix} + \begin{bmatrix} 3 & 1 & 2 \\ 2 & 1 & 3 \end{bmatrix} = \begin{bmatrix} 4 & 3 & 5 \\ 4 & 4 & 4 \end{bmatrix}.$$

Thus, we informally say that matrices are added entry-wise. Matrices of different sizes can never be added. It is easy to see that

$$A + B = B + A, \quad A + 0 = 0 + A = A$$

for all matrices $A, B \in \mathbb{F}^{m \times n}$, with an implicit understanding that $0 \in \mathbb{F}^{m \times n}$.

Similarly, matrices can be **multiplied by a scalar** entry-wise. If $\alpha \in \mathbb{F}$ and $A = [a_{ij}] \in \mathbb{F}^{m \times n}$, then

$$\alpha A = [\alpha a_{ij}] \in \mathbb{F}^{m \times n}.$$

Therefore, a scalar matrix with α on the diagonal is written as αI.

For $A = [a_{ij}]$, the matrix $-A \in \mathbb{F}^{m \times n}$ is taken as one whose (i, j)th entry is $-a_{ij}$. Thus,

$$-A = (-1)A, \quad (-A) + A = A + (-A) = 0.$$

We also abbreviate $A + (-B)$ to $A - B$, as usual. For example,

$$3 \begin{bmatrix} 1 & 2 & 3 \\ 2 & 3 & 1 \end{bmatrix} - \begin{bmatrix} 3 & 1 & 2 \\ 2 & 1 & 3 \end{bmatrix} = \begin{bmatrix} 0 & 5 & 7 \\ 4 & 8 & 0 \end{bmatrix}.$$

Addition and scalar multiplication of matrices satisfy the following properties: Let $A, B, C \in \mathbb{F}^{m \times n}$, and let $\alpha, \beta \in \mathbb{F}$. Then,

1. $A + B = B + A$.
2. $(A + B) + C = A + (B + C)$.
3. $A + 0 = 0 + A = A$.
4. $A + (-A) = (-A) + A = 0$.
5. $\alpha(\beta A) = (\alpha \beta) A$.
6. $\alpha(A + B) = \alpha A + \alpha B$.
7. $(\alpha + \beta)A = \alpha A + \beta A$.
8. $1 A = A$.
9. $(-1) A = -A$.

Notice that whatever we discuss here for matrices apply to row vectors and column vectors, in particular. But remember that a row vector cannot be added to a column vector unless both are of size 1×1.

We also define **multiplication** or product of matrices. Let $A = [a_{ik}] \in \mathbb{F}^{m \times n}$, and let $B = [b_{kj}] \in \mathbb{F}^{n \times r}$. Then, their **product** AB is a matrix $[c_{ij}] \in \mathbb{F}^{m \times r}$, where the (i, j)th entry is given by

$$c_{ij} = a_{i1}b_{1j} + \cdots + a_{in}b_{nj} = \sum_{k=1}^{n} a_{ik}b_{kj}.$$

Mark the sizes of A and B. The matrix product AB is defined only when the number of columns in A is equal to the number of rows in B. The result AB has number of rows as that of A and the number of columns as that of B.

A particular case might be helpful. Suppose u is a row vector in $\mathbb{F}^{1 \times n}$ and v is a column vector in $\mathbb{F}^{n \times 1}$. Then, their product $uv \in \mathbb{F}^{1 \times 1}$. It is a 1×1 matrix. Often, we identify such matrices with scalars. The product now looks like:

$$\begin{bmatrix} a_1 & \cdots & a_n \end{bmatrix} \begin{bmatrix} b_1 \\ \vdots \\ b_n \end{bmatrix} = [a_1 b_1 + \cdots + a_n b_n].$$

This is helpful in visualizing the general case, which looks like

$$
\begin{bmatrix} a_{11} & a_{1k} & a_{1n} \\ a_{i1} & \cdots & \mathbf{a_{ik}} & \cdots & \mathbf{a_{in}} \\ a_{m1} & a_{mk} & a_{mn} \end{bmatrix}
\begin{bmatrix} b_{11} & \mathbf{b_{1j}} & b_{1r} \\ \vdots \\ b_{\ell 1} & \mathbf{b_{\ell j}} & b_{\ell r} \\ \vdots \\ b_{n1} & \mathbf{b_{nj}} & b_{nr} \end{bmatrix}
=
\begin{bmatrix} c_{11} & c_{1j} & c_{1r} \\ c_{i1} & \mathbf{c_{ij}} & c_{ir} \\ c_{m1} & c_{mj} & c_{mr} \end{bmatrix}.
$$

The ith row of A multiplied with the jth column of B gives the (i, j)th entry in AB. Thus to get AB, you have to multiply all m rows of A with all r columns of B, taking one from each in turn. For example,

$$
\begin{bmatrix} 3 & 5 & -1 \\ 4 & 0 & 2 \\ -6 & -3 & 2 \end{bmatrix}
\begin{bmatrix} 2 & -2 & 3 & 1 \\ 5 & 0 & 7 & 8 \\ 9 & -4 & 1 & 1 \end{bmatrix}
=
\begin{bmatrix} 22 & -2 & 43 & 42 \\ 26 & -16 & 14 & 6 \\ -9 & 4 & -37 & -28 \end{bmatrix}.
$$

If $u \in \mathbb{F}^{1 \times n}$ and $v \in \mathbb{F}^{n \times 1}$, then $uv \in \mathbb{F}^{1 \times 1}$; but $vu \in \mathbb{F}^{n \times n}$. For instance,

$$
\begin{bmatrix} 3 & 6 & 1 \end{bmatrix} \begin{bmatrix} 1 \\ 2 \\ 4 \end{bmatrix} = \begin{bmatrix} 19 \end{bmatrix}, \quad
\begin{bmatrix} 1 \\ 2 \\ 4 \end{bmatrix} \begin{bmatrix} 3 & 6 & 1 \end{bmatrix} = \begin{bmatrix} 3 & 6 & 1 \\ 6 & 12 & 2 \\ 12 & 24 & 4 \end{bmatrix}.
$$

It shows clearly that matrix multiplication is not commutative. Commutativity can break down due to various reasons. First of all when AB is defined, BA may not be defined. Secondly, even when both AB and BA are defined, they may not be of the same size; thirdly, even when they are of the same size, they need not be equal. For example,

$$
\begin{bmatrix} 1 & 2 \\ 2 & 3 \end{bmatrix} \begin{bmatrix} 0 & 1 \\ 2 & 3 \end{bmatrix} = \begin{bmatrix} 4 & 7 \\ 6 & 11 \end{bmatrix} \quad \text{but} \quad
\begin{bmatrix} 0 & 1 \\ 2 & 3 \end{bmatrix} \begin{bmatrix} 1 & 2 \\ 2 & 3 \end{bmatrix} = \begin{bmatrix} 2 & 3 \\ 8 & 13 \end{bmatrix}.
$$

It does not mean that AB is never equal to BA. If $A, B \in \mathbb{F}^{m \times m}$ and A is a scalar matrix, then $AB = BA$. Conversely, if $A \in \mathbb{F}^{m \times m}$ is such that $AB = BA$ for all $B \in \mathbb{F}^{m \times m}$, then A must be a scalar matrix. This fact is not obvious, and its proof is involved. Moreover, there can be some particular non-scalar matrices A and B both in $\mathbb{F}^{n \times n}$ such that $AB = BA$.

Observe that if $A \in \mathbb{F}^{m \times n}$, then $AI_n = A$ and $I_m A = A$. Look at the columns of I_n in this product. They say that

$$
Ae_j = \text{the } j\text{th column of } A \quad \text{for } j = 1, \ldots, n.
$$

Here, e_j is the standard jth basis vector, the jth column of the identity matrix of order n; its jth component is 1, and all other components are 0. The above identity can also be seen by directly multiplying A with e_j, as in the following:

$$Ae_j = \begin{bmatrix} a_{11} & \cdots & a_{1j} & \cdots & a_{1n} \\ & & \vdots & & \\ a_{i1} & \cdots & a_{ij} & \cdots & a_{in} \\ & & \vdots & & \\ a_{m1} & \cdots & a_{mj} & \cdots & a_{mn} \end{bmatrix} \begin{bmatrix} 0 \\ \vdots \\ 1 \\ \vdots \\ 0 \end{bmatrix} = \begin{bmatrix} a_{1j} \\ \vdots \\ a_{ij} \\ \vdots \\ a_{mj} \end{bmatrix} = j\text{th column of } A.$$

Thus, A can be written in block form as

$$A = \begin{bmatrix} Ae_1 & \cdots & Ae_j & \cdots & Ae_n \end{bmatrix}.$$

Unlike numbers, product of two nonzero matrices can be a zero matrix. For instance,

$$\begin{bmatrix} 1 & 0 \\ 0 & 0 \end{bmatrix} \begin{bmatrix} 0 & 0 \\ 0 & 1 \end{bmatrix} = \begin{bmatrix} 0 & 0 \\ 0 & 0 \end{bmatrix}.$$

Let $A \in \mathbb{F}^{m \times n}$. We write its ith row as $A_{i\star}$ and its kth column as $A_{\star k}$.

We can now write A as a row of columns and also as a column of rows in the following manner:

$$A = [a_{ik}] = \begin{bmatrix} A_{\star 1} & \cdots & A_{\star n} \end{bmatrix} = \begin{bmatrix} A_{1\star} \\ \vdots \\ A_{m\star} \end{bmatrix}.$$

Write $B \in \mathbb{F}^{n \times r}$ similarly as

$$B = [b_{kj}] = \begin{bmatrix} B_{\star 1} & \cdots & B_{\star r} \end{bmatrix} = \begin{bmatrix} B_{1\star} \\ \vdots \\ B_{\star n} \end{bmatrix}.$$

Then, their product AB can now be written in block form as (ignoring extra brackets):

$$AB = \begin{bmatrix} AB_{\star 1} & \cdots & AB_{\star r} \end{bmatrix} = \begin{bmatrix} A_{1\star}B \\ \vdots \\ A_{m\star}B \end{bmatrix}.$$

It is easy to verify the following properties of matrix multiplication:

1. If $A \in \mathbb{F}^{m \times n}$, $B \in \mathbb{F}^{n \times r}$ and $C \in \mathbb{F}^{r \times p}$, then $(AB)C = A(BC)$.
2. If $A, B \in \mathbb{F}^{m \times n}$ and $C \in \mathbb{F}^{n \times r}$, then $(A + B)C = AB + AC$.
3. If $A \in \mathbb{F}^{m \times n}$ and $B, C \in \mathbb{F}^{n \times r}$, then $A(B + C) = AB + AC$.
4. If $\alpha \in \mathbb{F}$, $A \in \mathbb{F}^{m \times n}$ and $B \in \mathbb{F}^{n \times r}$, then $\alpha(AB) = (\alpha A)B = A(\alpha B)$.

Powers of square matrices can be defined inductively by taking

$$A^0 = I \quad \text{and} \quad A^n = AA^{n-1} \quad \text{for } n \in \mathbb{N}.$$

Example 1.1 Let $A = \begin{bmatrix} 1 & 1 & 0 \\ 0 & 1 & 2 \\ 0 & 0 & 1 \end{bmatrix}$. We show that $A^n = \begin{bmatrix} 1 & n & n(n-1) \\ 0 & 1 & 2n \\ 0 & 0 & 1 \end{bmatrix}$ for $n \in \mathbb{N}$.

We use induction on n. The basis case $n = 1$ is obvious. Suppose A^n is as given. Now,

$$A^{n+1} = AA^n = \begin{bmatrix} 1 & 1 & 0 \\ 0 & 1 & 2 \\ 0 & 0 & 1 \end{bmatrix} \begin{bmatrix} 1 & n & n(n-1) \\ 0 & 1 & 2n \\ 0 & 0 & 1 \end{bmatrix} = \begin{bmatrix} 1 & n+1 & (n+1)n \\ 0 & 1 & 2(n+1) \\ 0 & 0 & 1 \end{bmatrix}.$$

By convention, $A^0 = I$. Incidentally, taking $n = 0$ in A^n, we have $A^0 = I$. $\qquad\square$

A square matrix A of order m is called **invertible** iff there exists a matrix B of order m such that

$$AB = I = BA.$$

If B and C are matrices that satisfy the above equations, then

$$C = CI = C(AB) = (CA)B = IB = B.$$

Therefore, there exists a unique matrix B corresponding to any given invertible matrix A such that $AB = I = BA$. We write such a matrix B as A^{-1} and call it the **inverse** of the matrix A. That is, A^{-1} is that matrix which satisfies

$$AA^{-1} = I = A^{-1}A.$$

We talk of invertibility of square matrices only; all square matrices are not invertible. For example, I is invertible but 0 is not. If $AB = 0$ for nonzero square matrices A and B, then neither A nor B is invertible. Why?

If both $A, B \in \mathbb{F}^{n \times n}$ are invertible, then $(AB)^{-1} = B^{-1}A^{-1}$. Reason:

$$B^{-1}A^{-1}AB = B^{-1}IB = I = AIA^{-1} = ABB^{-1}A^{-1}.$$

Invertible matrices play a crucial role in solving linear systems uniquely. We will come back to the issue later.

Exercises for Sect. 1.2

1. Compute $AB, CA, DC, DCAB, A^2, D^2$ and A^3B^2, where

$$A = \begin{bmatrix} 2 & 3 \\ 1 & 2 \end{bmatrix}, \quad B = \begin{bmatrix} 4 & -1 \\ 4 & 0 \end{bmatrix}, \quad C = \begin{bmatrix} -1 & 2 \\ 2 & -1 \\ 1 & 3 \end{bmatrix}, \quad D = \begin{bmatrix} 3 & 2 & 1 \\ 4 & -6 & 0 \\ 1 & -2 & -2 \end{bmatrix}.$$

2. Let $A = [a_{ij}] \in \mathbb{F}^{2\times2}$ with $a_{11} \neq 0$. Let $b = a_{21}/a_{11}$. Show that there exists $c \in \mathbb{F}$ such that $A = \begin{bmatrix} 1 & 0 \\ b & 1 \end{bmatrix} \begin{bmatrix} a_{11} & a_{12} \\ 0 & c \end{bmatrix}$. What could be c?

3. Let $A \in \mathbb{F}^{m\times n}$, and let $B \in \mathbb{F}^{n\times k}$. For $1 \leq i \leq m, 1 \leq j \leq k$, show that
   (a) $(AB)_{i\star} = A_{i\star} B$ (b) $(AB)_{\star j} = A B_{\star j}$

4. Construct two 3×3 matrices A and B such that $AB = 0$ but $BA \neq 0$.

5. Can you construct invertible 2×2 matrices A and B such that $AB = 0$?

6. Let $A, B \in \mathbb{F}^{n\times n}$ be such that $AB = 0$. Then which of the following is/are true, and why?

   (a) At least one of A or B is the zero matrix.
   (b) At least one of A or B is invertible.
   (c) At least one of A or B is non-invertible.
   (d) If $A \neq 0$ and $B \neq 0$, then neither is invertible.

7. Prove all properties of multiplication of matrices mentioned in the text.

8. Let A be the 4×4 matrix whose super-diagonal entries are all 1, and all other entries 0. Show that $A^n = 0$ for $n \geq 4$.

9. Let A be the 4×4 matrix with each diagonal entry as $\frac{1}{2}$, and each non-diagonal entry as $-\frac{1}{2}$. Compute A^n for $n \in \mathbb{N}$.

1.3 Transpose and Adjoint

Given a matrix $A \in \mathbb{F}^{m\times n}$, its **transpose** is the matrix in $\mathbb{F}^{n\times m}$ defined by

$$\text{the } (i, j)\text{th entry of } A^t = \text{the } (j, i)\text{th entry of } A.$$

The transpose of A is denoted by A^t.

That is, the ith column of A^t is the column vector $[a_{i1}, \ldots, a_{in}]^t$. In this sense, the rows of A become the columns of A^t and the columns of A become the rows of A^t. For example,

$$\text{if } A = \begin{bmatrix} 1 & 2 & 3 \\ 2 & 3 & 1 \end{bmatrix} \text{ then } A^t = \begin{bmatrix} 1 & 2 \\ 2 & 3 \\ 3 & 1 \end{bmatrix} \text{ and } (A^t)^t = \begin{bmatrix} 1 & 2 & 3 \\ 2 & 3 & 1 \end{bmatrix}.$$

In particular, if $u = [a_1, \ldots a_m]$ is a row vector, then its transpose is

$$u^t = \begin{bmatrix} a_1 \\ \vdots \\ a_m \end{bmatrix},$$

which is a column vector, as mentioned earlier. Similarly, the transpose of a column vector is a row vector. If you write A as a row of column vectors, then you can express A^t as a column of row vectors.

$$\text{For } A = \begin{bmatrix} A_{\star 1} & \cdots & A_{\star n} \end{bmatrix} = \begin{bmatrix} A_{1\star} \\ \vdots \\ A_{m\star} \end{bmatrix}, \quad A^t = \begin{bmatrix} A^t_{\star 1} \\ \vdots \\ A^t_{\star n} \end{bmatrix} = \begin{bmatrix} A^t_{1\star} & \cdots & A^t_{m\star} \end{bmatrix}.$$

The following are some of the properties of the operation of transpose.

1. $(A^t)^t = A$.
2. $(A + B)^t = A^t + B^t$.
3. $(\alpha A)^t = \alpha A^t$.
4. $(AB)^t = B^t A^t$.
5. If A is invertible, then A^t is invertible, and $(A^t)^{-1} = (A^{-1})^t$.

In the above properties, we assume that the operations are allowed; that is, in (2), A and B must be matrices of the same size. In (3), α is a scalar. Similarly, in (4), the number of columns in A must be equal to the number of rows in B; and in (5), A must be a square matrix.

It is easy to see all the above properties, except perhaps the fourth one. For this, let $A \in \mathbb{F}^{m \times n}$ and let $B \in \mathbb{F}^{n \times r}$. Now, the (j, i)th entry in $(AB)^t$ is the (i, j)th entry in AB; it is given by

$$a_{i1}b_{j1} + \cdots + a_{in}b_{jn}.$$

On the other side, the (j, i)th entry in $B^t A^t$ is obtained by multiplying the jth row of B^t with the ith column of A^t. This is same as multiplying the entries in the jth column of B with the corresponding entries in the ith row of A and then taking the sum. Thus, it is

$$b_{j1}a_{i1} + \cdots + b_{jn}a_{in}.$$

This is the same as computed earlier.

The fifth one follows from the fourth one and the fact that $(AB)^{-1} = B^{-1}A^{-1}$.

Observe that transpose of a lower triangular matrix is an upper triangular matrix and vice versa.

Close to the operations of transpose of a matrix is the adjoint. Recall that the **complex conjugate** of a scalar α is written as $\overline{\alpha}$, where $\overline{a + ib} = a - ib$ for $a, b \in \mathbb{R}$. The matrix obtained from A by taking complex conjugate of each entry is written as $\overline{A}$. That is, the (i, j)th entry of $\overline{A}$ is the complex conjugate of the (i, j)th entry of A. The **adjoint** of a matrix $A = [a_{ij}] \in \mathbb{F}^{m \times n}$, written as A^*, is the transpose of $\overline{A}$. That is, $A^* = (\overline{A})^t = \overline{A^t}$. The adjoint of A is also called the **conjugate transpose** of A. It may be defined directly by

the (i, j)th entry of $A^* =$ the complex conjugate of (j, i)th entry of A.

For instance, if $A = \begin{bmatrix} 1+i & 2 & 3 \\ 2 & 3 & 1-i \end{bmatrix}$ then $A^* = \begin{bmatrix} 1-i & 2 \\ 2 & 3 \\ 3 & 1+i \end{bmatrix}$.

If $A = [a_{ij}] \in \mathbb{R}^{m \times n}$, then $\bar{a}_{ij} = a_{ij}$. Consequently, $A^* = A^t$. For example,

$$\text{if } A = \begin{bmatrix} 1 & 2 & 3 \\ 2 & 3 & 1 \end{bmatrix} \text{ then } A^* = \begin{bmatrix} 1 & 2 \\ 2 & 3 \\ 3 & 1 \end{bmatrix} = A^t.$$

The ith column of A^* is the column vector $[\bar{a}_{i1}, \ldots, \bar{a}_{in}]^t$, which is the adjoint of the ith row of A. We may write the operation of adjoint in terms of rows and columns of a matrix as in the following.

$$\text{For } A = \begin{bmatrix} A_{\star 1} & \cdots & A_{\star n} \end{bmatrix} = \begin{bmatrix} A_{1\star} \\ \vdots \\ A_{m\star} \end{bmatrix}, \quad A^* = \begin{bmatrix} A^*_{\star 1} \\ \vdots \\ A^*_{\star n} \end{bmatrix} = \begin{bmatrix} A^*_{1\star} & \cdots & A^*_{m\star} \end{bmatrix}.$$

The operation of adjoint satisfies the following properties:

1. $(A^*)^* = A$.
2. $(A + B)^* = A^* + B^*$.
3. $(\alpha A)^* = \bar{\alpha} A^*$.
4. $(AB)^* = B^* A^*$.
5. If A is invertible, then A^* is invertible, and $(A^*)^{-1} = (A^{-1})^*$.

In (2), the matrices A and B must be of the same size, and in (4), the number of columns in A must be equal to the number of rows in B.

Exercises for Sect. 1.3

1. Determine $A^t, \bar{A}, A^*, A^*A$ and AA^*, where

(a) $A = \begin{bmatrix} -1 & 2 & 3 & 1 \\ 2 & -1 & 0 & 3 \\ 0 & -1 & -3 & 1 \end{bmatrix}$ (b) $A = \begin{bmatrix} 1 & -2+i & 3-i \\ i & -1-i & 2i \\ 1+3i & -i & -3 \\ -2 & 0 & -i \end{bmatrix}$

2. Prove all properties of the transpose and the adjoint mentioned in the text.
3. Let $A \in \mathbb{C}^{m \times n}$. Suppose $AA^* = I_m$. Does it follow that $A^*A = I_n$?

1.4 Elementary Row Operations

Recall that while solving linear equations in two or three variables, we try to eliminate a variable from all but one equation by adding an equation to the other, or even adding a constant times one equation to another. We do similar operations on the rows of a matrix. Theoretically, it will be advantageous to see these operations as multiplication by some special kinds of matrices.

Let $e_1, \ldots, e_m$ be the standard basis vectors of $\mathbb{F}^{m \times 1}$. Let $1 \leq i, j \leq m$. The product $e_i e_j^t$ is an $m \times m$ matrix whose (i, j)th entry is 1, and all other entries are 0. We write such a matrix as E_{ij}. For instance, when $m = 3$, we have

$$e_2 e_3^t = \begin{bmatrix} 0 \\ 1 \\ 0 \end{bmatrix} \begin{bmatrix} 0 & 0 & 1 \end{bmatrix} = \begin{bmatrix} 0 & 0 & 0 \\ 0 & 0 & 1 \\ 0 & 0 & 0 \end{bmatrix} = E_{23}.$$

An **elementary matrix** of order m is one of the following three types of matrices:

1. $E[i, j] = I - E_{ii} - E_{jj} + E_{ij} + E_{ji}$ with $i \neq j$.
2. $E_\alpha[i] = I - E_{ii} + \alpha E_{ii}$, where α is a nonzero scalar.
3. $E_\alpha[i, j] = I + \alpha E_{ij}$, where α is a nonzero scalar and $i \neq j$.

Here, I is the identity matrix of order m. Similarly, the order of the elementary matrices will be understood from the context; we will not show that in our symbolism.

The following are instances of elementary matrices of order 3:

$$E[1, 2] = \begin{bmatrix} 0 & 1 & 0 \\ 1 & 0 & 0 \\ 0 & 0 & 1 \end{bmatrix}, \quad E_{-1}[2] = \begin{bmatrix} 1 & 0 & 0 \\ 0 & -1 & 0 \\ 0 & 0 & 1 \end{bmatrix}, \quad E_2[3, 1] = \begin{bmatrix} 1 & 0 & 0 \\ 0 & 1 & 0 \\ 2 & 0 & 1 \end{bmatrix}.$$

We see that $E_{ij} A = e_i e_j^t A = e_i (A^t e_j)^t = e_i ((A^t)_{\star j})^t = e_i A_{j\star}$. The last matrix has all rows as zero rows except the ith one, which is equal to $A_{j\star}$, the jth row of A. That is,

$E_{ij} A$ is an $m \times n$ matrix whose ith row is the jth row of A, and the rest of its rows are zero rows. In particular, $E_{ii} A$ is the matrix obtained from A by replacing all its rows by zero rows except the ith one.

Using these, we observe the following.

Observation 1.1 1. $E[i, j] A$ is obtained from A by exchanging its ith and jth rows.
2. $E_\alpha[i] A$ is obtained from A by multiplying its ith row with α.
3. $E_\alpha[i, j] A$ is obtained from A by replacing its ith row with the ith row plus α times the jth row.

We call each of these operations of pre-multiplying a matrix with an elementary matrix as an **elementary row operation**. Thus, there are three kinds of elementary row operations corresponding to those listed in Observation 1.1. Sometimes, we will refer to them as of Type 1, 2, or 3, respectively. Also, in computations, we will write

$$A \xrightarrow{E} B$$

to mean that the matrix B has been obtained from A by an elementary row operation E, that is, when $B = EA$.

Example 1.2 Verify the following applications of elementary row operations:

$$
\begin{bmatrix} 1 & 1 & 1 \\ 2 & 2 & 2 \\ 3 & 3 & 3 \end{bmatrix}
\xrightarrow{E_{-3}[3,1]}
\begin{bmatrix} 1 & 1 & 1 \\ 2 & 2 & 2 \\ 0 & 0 & 0 \end{bmatrix}
\xrightarrow{E_{-2}[2,1]}
\begin{bmatrix} 1 & 1 & 1 \\ 0 & 0 & 0 \\ 0 & 0 & 0 \end{bmatrix}.
\qquad \square
$$

Often, we will apply elementary row operations in a sequence. In this way, the above operations could be shown in one step as $E_{-3}[3, 1]$, $E_{-2}[2, 1]$. However, remember that the result of application of this sequence of elementary row operations on a matrix A is $E_{-2}[2, 1] E_{-3}[3, 1] A$; the products are in reverse order.

Observe that each elementary matrix is invertible. In fact, $E[i, j]$ is its own inverse, $E_{1/\alpha}[i]$ is the inverse of $E_{\alpha}[i]$, and $E_{-\alpha}[i, j]$ is the inverse of $E_{\alpha}[i, j]$. Therefore, any product of elementary matrices is invertible. It follows that if B has been obtained from A by applying a sequence of elementary row operations, then A can also be obtained from B by a sequence of elementary row operations.

Exercises for Sect. 1.4

1. Show the following:

   (a) $E_{ij} E_{jm} = E_{im}$; if $j \neq k$, then $E_{ij} E_{km} = 0$.
   (b) Each $A = [a_{ij}] \in \mathbb{F}^{n \times n}$ can be written as $A = \sum_{i=1}^{n} \sum_{j=1}^{n} a_{ij} E_{ij}$.

2. Compute $E[2, 3]A$, $E_i[2]A$, $E_{-1/2}[1, 3]A$ and $E_i[1, 2]A$, where

   (a) $A = \begin{bmatrix} -1 & 2 & 3 & 1 \\ 2 & -1 & 0 & 3 \\ 0 & -1 & -3 & 1 \end{bmatrix}$ (b) $A = \begin{bmatrix} 1 & -2+i & 3-i \\ i & -1-i & 2i \\ 1+3i & -i & -3 \\ -2 & 0 & -i \end{bmatrix}$

3. Take an invertible 2×2 matrix. Bring it to the identity matrix of order 2 by applying elementary row operations.

4. Take a non-invertible 2×2 matrix. Try to bring it to the identity matrix of order 2 by applying elementary row operations.

5. Argue in general terms why the following in Observation 1.1 are true:

   (a) $E[i, j] A$ is obtained from A by exchanging its ith and jth rows.
   (b) $E_{\alpha}[i] A$ is obtained from A by multiplying its ith row with α.
   (c) $E_{\alpha}[i, j] A$ is obtained from A by adding to its ith row α times the jth row.

6. How can the elementary matrices be obtained from the identity matrix?

7. Let α be a nonzero scalar. Show the following:

   (a) $(E[i, j])^t = E[i, j]$, $(E_{\alpha}[i])^t = E_{\alpha}[i]$, $(E_{\alpha}[i, j])^t = E[j, i]$.
   (b) $(E[i, j])^{-1} = E[i, j]$, $(E_{\alpha}[i])^{-1} = E_{1/\alpha}[i]$, $(E_{\alpha}[i, j])^{-1} = E_{-\alpha}[i, j]$.

8. For each of the following pairs of matrices, find an elementary matrix E such that $B = EA$.

   (a) $A = \begin{bmatrix} 2 & 1 & 3 \\ 3 & 1 & 4 \\ -2 & 4 & 5 \end{bmatrix}$, $B = \begin{bmatrix} 2 & 1 & 3 \\ -2 & 4 & 5 \\ 1 & 5 & 9 \end{bmatrix}$

(b) $A = \begin{bmatrix} 4 & -2 & 3 \\ 1 & 0 & 2 \\ 0 & 3 & 5 \end{bmatrix}$, $B = \begin{bmatrix} 3 & -2 & 1 \\ 1 & 0 & 2 \\ 0 & 3 & 5 \end{bmatrix}$

9. For each of the following pairs of matrices, find an elementary matrix E such that $B = AE$. [Hint: The requirement is $B^t = E^t A^t$.]

(a) $A = \begin{bmatrix} 3 & 1 & 4 \\ 4 & 1 & 2 \\ 2 & 3 & 1 \end{bmatrix}$, $B = \begin{bmatrix} 4 & 5 & 3 \\ 2 & 3 & 4 \\ 1 & 4 & 2 \end{bmatrix}$

(b) $A = \begin{bmatrix} 2 & -2 & 3 \\ -1 & 4 & 2 \\ 3 & 1 & -2 \end{bmatrix}$, $B = \begin{bmatrix} 4 & -2 & 1 \\ -2 & 4 & -4 \\ 6 & 1 & 8 \end{bmatrix}$

1.5 Row Reduced Echelon Form

Consider solving the linear equations $x + y = 3$ and $-x + y = 1$. If we add the equations, we obtain $2y = 4$ or $y = 2$. Substituting this value in one of the equations, we get $x = 1$. In terms of matrices, we write the equations as

$$\begin{bmatrix} 1 & 1 & 3 \\ -1 & 1 & 1 \end{bmatrix}$$

where the third column is the right-hand side of the equality sign in our equations. Our method of solution suggests that we proceed as follows:

$$\begin{bmatrix} 1 & 1 & 3 \\ -1 & 1 & 1 \end{bmatrix} \xrightarrow{E_1[2,1]} \begin{bmatrix} 1 & 1 & 3 \\ 0 & 2 & 4 \end{bmatrix} \xrightarrow{E_{1/2}[2]} \begin{bmatrix} 1 & 1 & 3 \\ 0 & 1 & 2 \end{bmatrix} \xrightarrow{E_{-1}[1,2]} \begin{bmatrix} 1 & 0 & 1 \\ 0 & 1 & 2 \end{bmatrix}$$

In the final result, the first 2×2 block is an identity matrix. Thus, we could obtain a unique solution as $x = 1$, $y = 2$, which are the respective entries in the last column.

As you have seen, we may not be able to bring any arbitrary square matrix to the identity matrix of the same order by elementary row operations. On the other hand, if two rows of a matrix are same, then one of them can be made a zero row after a suitable elementary row operation. We may thus look for a matrix with as many zero entries as possible and somewhat closer to the identity matrix. Moreover, such a matrix if invertible must be the identity matrix. We would like to define such a matrix looking at our requirements on the end result.

Recall that in a matrix, the row index of the ith row is i, which is also called the row index of the (i, j)th entry. Similarly, j is the column index of the jth column and also of the (i, j)th entry.

In a nonzero row of a matrix, the nonzero entry with minimum column index (first from left) is called a **pivot**. We mark a pivot by putting a box around it. A column where a pivot occurs is called a **pivotal column**.

In the following matrix, the pivots are shown in boxes:

$$\begin{bmatrix} 0 & \boxed{1} & 2 & 0 \\ 0 & 0 & 0 & 0 \\ 0 & 0 & 0 & \boxed{2} \end{bmatrix}$$

The row index of the pivot 1 is 1, and its column index is 2. The row index of the pivot 2 is 3, and its column index is 4. The column indices of the pivotal columns are 2 and 4.

A matrix $A \in \mathbb{F}^{m \times n}$ is said to be in **row reduced echelon form** (RREF) iff the following conditions are satisfied:

1. Each pivot is equal to 1.
2. In a pivotal column, all entries other than the pivot are zero.
3. The row index of each nonzero row is smaller than the row index of each zero row.
4. Among any two pivots, the pivot with larger row index also has larger column index.

Example 1.3 The following matrices are in row reduced echelon form:

$$\begin{bmatrix} \boxed{1} & 2 & 0 & 0 \\ 0 & 0 & \boxed{1} & 0 \\ 0 & 0 & 0 & \boxed{1} \end{bmatrix}, \begin{bmatrix} 0 \\ 0 \\ 0 \\ 0 \end{bmatrix}, \begin{bmatrix} \boxed{1} \\ 0 \\ 0 \\ 0 \end{bmatrix}, [0\ 0\ 0\ 0], [0\ \boxed{1}\ 0\ 2].$$

The following are not in row reduced echelon form:

$$\begin{bmatrix} 0 & \boxed{1} & 3 & 0 \\ 0 & 0 & 0 & \boxed{2} \\ 0 & 0 & 0 & 0 \\ 0 & 0 & 0 & 0 \end{bmatrix}, \begin{bmatrix} 0 & \boxed{1} & 3 & 1 \\ 0 & 0 & 0 & \boxed{1} \\ 0 & 0 & 0 & 0 \\ 0 & 0 & 0 & 0 \end{bmatrix}, \begin{bmatrix} 0 & \boxed{1} & 0 & 0 \\ 0 & 0 & \boxed{1} & 0 \\ 0 & 0 & 0 & 0 \\ 0 & 0 & 0 & \boxed{1} \end{bmatrix}, \begin{bmatrix} 0 & \boxed{1} & 3 & 0 \\ 0 & 0 & 0 & \boxed{1} \\ 0 & 0 & 0 & \boxed{1} \\ 0 & 0 & 0 & 0 \end{bmatrix}. \qquad \square$$

In a row reduced echelon form matrix, all zero rows (if there are any) occur at the bottom of the matrix. Further, the pivot in a latter row occurs below and to the right of any former row.

A column vector (an $n \times 1$ matrix) in row reduced echelon form is either the zero vector or e_1.

If a matrix in RREF has k pivotal columns, then those columns occur in the matrix as $e_1, \ldots, e_k$, read from left to right, though there can be other columns between these pivotal columns.

Further, if a non-pivotal column occurs between two pivotal columns e_k and e_{k+1}, then the entries of the non-pivotal column beyond the kth entry are all zero.

In a row reduced echelon form matrix, all entries below and to the left of any pivot are zero. Ignoring such zero entries and drawing lines below and to the left of pivots, a pattern of steps emerges, thus the name *echelon form*.

Any matrix can be brought to a row reduced echelon form by using elementary row operations. We first search for a pivot and make it 1; then using elementary row operations, we zero-out all entries except the pivot in a column and then use row exchanges to take the zero rows to the bottom. Following these guidelines, we give an algorithm to reduce a matrix to its RREF.

Reduction to Row Reduced Echelon Form

1. Set the work region R to the whole matrix A.
2. If all entries in R are 0, then stop.
3. If there are nonzero entries in R, then find the leftmost nonzero column. Mark it as the pivotal column.
4. Find the topmost nonzero entry in the pivotal column. Box it; it is a pivot.
5. If the pivot is not on the top row of R, then exchange the row of A which contains the top row of R with the row where the pivot is.
6. If the pivot, say, α is not equal to 1, then replace the top row of R in A by $1/\alpha$ times that row. Mark the top row of R in A as the *pivotal row*.
7. Zero-out all entries, except the pivot, in the pivotal column by replacing each row above and below the pivotal row using an elementary row operation (of Type 3) in A with that row and the pivotal row.
8. If the pivot is the rightmost and the bottommost entry in A, then stop. Else, find the sub-matrix to the right and below the pivot. Reset the work region R to this sub-matrix, and go to Step 2.

We will refer to the output of the above reduction algorithm as *the row reduced echelon form* (**the RREF**) of a given matrix.

Example 1.4 $A = \begin{bmatrix} \boxed{1} & 1 & 2 & 0 \\ 3 & 5 & 7 & 1 \\ 1 & 5 & 4 & 5 \\ 2 & 8 & 7 & 9 \end{bmatrix} \xrightarrow{R1} \begin{bmatrix} \boxed{1} & 1 & 2 & 0 \\ 0 & 2 & 1 & 1 \\ 0 & 4 & 2 & 5 \\ 0 & 6 & 3 & 9 \end{bmatrix} \xrightarrow{E_{1/2}[2]} \begin{bmatrix} \boxed{1} & 1 & 2 & 0 \\ 0 & \boxed{1} & 1/2 & 1/2 \\ 0 & 4 & 2 & 5 \\ 0 & 6 & 3 & 9 \end{bmatrix}$

$\xrightarrow{R2} \begin{bmatrix} \boxed{1} & 0 & 3/2 & -1/2 \\ 0 & \boxed{1} & 1/2 & 1/2 \\ 0 & 0 & 0 & 3 \\ 0 & 0 & 0 & 6 \end{bmatrix} \xrightarrow{E_{1/3}[3]} \begin{bmatrix} \boxed{1} & 0 & 3/2 & -1/2 \\ 0 & \boxed{1} & 1/2 & 1/2 \\ 0 & 0 & 0 & \boxed{1} \\ 0 & 0 & 0 & 6 \end{bmatrix} \xrightarrow{R3} \begin{bmatrix} \boxed{1} & 0 & 3/2 & 0 \\ 0 & \boxed{1} & 1/2 & 0 \\ 0 & 0 & 0 & \boxed{1} \\ 0 & 0 & 0 & 0 \end{bmatrix} = B.$

Here, $R1 = E_{-3}[2, 1]$, $E_{-1}[3, 1]$, $E_{-2}[4, 1]$, $R2 = E_{-1}[2, 1]$, $E_{-4}[3, 2]$, $E_{-6}[4, 2]$ and $R3 = E_{1/2}[1, 3]$, $E_{-1/2}[2, 3]$, $E_{-6}[4, 3]$. Notice that the matrix B, which is the RREF of A, is given by

$$B = E_{-6}[4, 3] \, E_{-1/2}[2, 3] \, E_{1/2}[1, 3] \, E_{1/3}[3] \, E_{-6}[4, 2] \, E_{-4}[3, 2]$$
$$E_{-1}[2, 1] \, E_{1/2}[2] \, E_{-2}[4, 1] \, E_{-1}[3, 1] \, E_{-3}[2, 1] \, A. \qquad \square$$

The number of pivots in the RREF of a matrix A is called the **rank** of the matrix; we denote it by rank(A).

Let $A \in \mathbb{F}^{m \times n}$. Suppose A has the columns as $u_1, \ldots, u_n$; these are column vectors from $\mathbb{F}^{m \times 1}$. Thus, we write $A = \begin{bmatrix} u_1 & \cdots & u_n \end{bmatrix}$.

Let B be the RREF of A obtained by applying a sequence of elementary row operations. Let E be the $m \times m$ invertible matrix, which is the product of the corresponding elementary matrices, so that

$$EA = E\begin{bmatrix} u_1 & \cdots & u_n \end{bmatrix} = B.$$

Suppose $\text{rank}(A) = r$. Then, the standard basis vectors $e_1, \ldots, e_r$ of $\mathbb{F}^{m \times 1}$ occur as the pivotal columns in B, in that order. Denote the $n - r$ non-pivotal columns in B as $v_1, \ldots, v_{n-r}$, occurring in B in that order.

Observation 1.2 In B, if v_i occurs between the pivotal columns e_j and e_{j+1}, then $v_i = [a_1, \ldots a_j, 0, 0, \ldots, 0]^t = a_1 e_1 + \cdots + a_j e_j$ for some $a_1, \ldots, a_j \in \mathbb{F}$.

Notice that the scalars $a_1, \ldots, a_j$ in the above observation need not all be nonzero.

If e_1 occurs as k_1th column in B, e_2 occurs as the k_2th column in B, and so on, then $E u_{k_1} = e_1$, $E u_{k_2} = e_2$, $\ldots$. In the notation of Observation 1.2,

$$v_i = a_1 E u_{k_1} + \cdots + a_j E u_{k_j}.$$

Then, $E^{-1} v_i = a_1 u_{k_1} + \cdots + a_j u_{k_j}$. As $E^{-1} v_i$ is the corresponding column of A, we observe the following.

Observation 1.3 In B, if a vector $v_i = [a_1, \ldots, a_j, 0, 0, \ldots, 0]^t$ occurs as the kth column, and prior to it occur the standard basis vectors $e_1, \ldots, e_j$ (and no other) in the columns $k_1, \ldots, k_j$, respectively, then $u_k = a_1 u_{k_1} + \cdots + a_j u_{k_j}$.

In Example 1.4, the first two columns appear as pivotal columns and the third one is a non-pivotal column in the RREF. Observation 1.2 says that the third column of A is some scalar times the first column plus a scalar times the second column; the scalars are precisely the entries in the third column of B. That is,

$$\begin{bmatrix} 2 \\ 7 \\ 4 \\ 7 \end{bmatrix} = \frac{3}{2} \begin{bmatrix} 1 \\ 3 \\ 1 \\ 2 \end{bmatrix} + \frac{1}{2} \begin{bmatrix} 1 \\ 5 \\ 5 \\ 8 \end{bmatrix}.$$

Notice that $\beta_1 u_{k_1} + \cdots + \beta_r u_{k_r} = \beta_1 E^{-1} e_1 + \cdots + \beta_r E^{-1} e_r$. Since e_{r+1} is not expressible in the form $\alpha_1 e_1 + \cdots + \alpha_r e_r$, we see that $E^{-1} e_{r+1}$ is not expressible in the form $\alpha_1 E^{-1} e_1 + \cdots + \alpha_r E^{-1} e_r$. Therefore, we observe the following:

Observation 1.4 If $\text{rank}(A) = r < m$, then $E^{-1} e_{r+1}, \ldots, E^{-1} e_m$ are not expressible in the form $\alpha_1 u_1 + \cdots + \alpha_n u_n$ for any $\alpha_1, \ldots, \alpha_n \in \mathbb{F}$.

Given a matrix A, our algorithm produces a unique matrix B in RREF. It raises the question whether by following another algorithm that employs elementary row operations on A, we would obtain a matrix in RREF different from B? The following result shows that this is not possible; the row reduced echelon form of a matrix is canonical.

Theorem 1.1 *Let $A \in \mathbb{F}^{m \times n}$. There exists a unique matrix in $\mathbb{F}^{m \times n}$ in row reduced echelon form obtained from A by elementary row operations.*

Proof Suppose $B, C \in \mathbb{F}^{m \times n}$ are matrices in RREF such that each has been obtained from A by elementary row operations. Since elementary matrices are invertible, $B = E_1 A$ and $C = E_2 A$ for some invertible matrices $E_1, E_2 \in \mathbb{F}^{m \times m}$. Now, $B = E_1 A = E_1 (E_2)^{-1} C$. Write $E = E_1 (E_2)^{-1}$ to have $B = EC$, where E is invertible.

Assume, on the contrary, that $B \neq C$. Then, there exists a column index, say $k \geq 1$, such that the first $k - 1$ columns of B coincide with the first $k - 1$ columns of C, respectively; and the kth column of B is not equal to the kth column of C. Let u be the kth column of B, and let v be the kth column of C. We have $u = Ev$ and $u \neq v$.

Suppose the pivotal columns that appear within the first $k - 1$ columns in C are $e_1, \ldots, e_j$. Then, $e_1, \ldots, e_j$ are also the pivotal columns in B that appear within the first $k - 1$ columns. Since $B = EC$, we have $C = E^{-1} B$; consequently,

$$e_1 = Ee_1 = E^{-1} e_1, \ldots, e_j = Ee_j = E^{-1} e_j.$$

Since C is in RREF, either $u = e_{j+1}$ or there exist scalars $\alpha_1, \ldots, \alpha_j$ such that $u = \alpha_1 e_1 + \cdots + \alpha_j e_j$. The latter case includes the possibility that $u = 0$. Similarly, either $v = e_{j+1}$ or $v = \beta_1 e_1 + \cdots + \beta_j e_j$ for some scalars $\beta_1, \ldots, \beta_j$. We consider the following exhaustive cases.

If $u = e_{j+1}$ and $v = e_{j+1}$, then $u = v$.
If $u = e_{j+1}$ and $v = \beta_1 e_1 + \cdots + \beta_j e_j$, then

$$u = Ev = \beta_1 Ee_1 + \cdots + \beta_j Ee_j = \beta_1 e_1 + \cdots + \beta_j e_j = v.$$

If $u = \alpha_1 e_1 + \cdots + \alpha_j u_j$ (and whether $v = e_{j+1}$ or $v = \beta_1 e_1 + \cdots + \beta_j e_j$), then

$$v = E^{-1} u = \alpha_1 E^{-1} e_1 + \cdots + \alpha_j E^{-1} e_j = \alpha_1 e_1 + \cdots + \alpha_j e_j = u.$$

In either case, $u = v$; this is a contradiction. Therefore, $B = C$. ∎

Theorem 1.1 justifies our use of the term *the* RREF of a matrix. Thus, the rank of a matrix does not depend on which algorithm we have followed in reducing it to its RREF.

Exercises for Sect. 1.5

1. Which of the following matrices are in RREF, and which are not?

(a) $\begin{bmatrix} 1 & 2 & 1 & 0 \\ 0 & 0 & 1 & 3 \\ 0 & 0 & 0 & 0 \end{bmatrix}$ (b) $\begin{bmatrix} 1 & 2 & 0 & 0 \\ 0 & 0 & 1 & 3 \\ 0 & 0 & 0 & 0 \end{bmatrix}$ (c) $\begin{bmatrix} 1 & 0 & 0 & 1 \\ 0 & 1 & 0 & 1 \\ 0 & 0 & 1 & 1 \end{bmatrix}$

2. Compute the RREF of the following matrices:

(a) $\begin{bmatrix} 0 & 0 & 1 \\ 0 & 1 & 0 \\ 1 & 0 & 0 \end{bmatrix}$ (b) $\begin{bmatrix} 2 & -1 & -1 & 0 \\ 1 & 0 & 1 & -4 \\ 0 & 1 & -1 & -4 \end{bmatrix}$ (c) $\begin{bmatrix} 1 & 2 & 1 & -1 \\ 0 & 2 & 3 & 3 \\ 1 & -1 & -3 & -4 \\ 1 & 1 & 5 & -2 \end{bmatrix}$

3. In Example 1.4, let u_i be the ith column of A, and let w_j be the jth row of A. Let $B = EA$ be the RREF of A.

   (a) Verify that $Eu_2 = e_2$.
   (b) Find $a, b \in \mathbb{R}$ such that $u_3 = au_1 + bu_2$ using the RREF of A.
   (c) Determine $a, b, c \in \mathbb{R}$ such that $w_4 = aw_1 + bw_2 + cw_3$.

4. Show that if an $n \times n$ matrix in RREF has rank n, then it is the identity matrix.

5. Suppose the RREF of a matrix A is equal to $\begin{bmatrix} 1 & 2 & 0 & 3 & 1 & -2 \\ 0 & 0 & 1 & 2 & 4 & 5 \\ 0 & 0 & 0 & 0 & 0 & 0 \\ 0 & 0 & 0 & 0 & 0 & 0 \end{bmatrix}$.

   If $A_{\star 1} = [1, 1, 3, 4]^t$ and $A_{\star 3} = [2, -1, 1, 3]^t$, then determine $A_{\star 6}$.

6. Consider the row vectors $v_1 = [1, 2, 3, 4]$, $v_2 = [2, 0, 1, 1]$, $v_3 = [-3, 2, 1, 2]$, and $v_4 = [1, -2, -2, -3]$. Construct a row vector $v \in \mathbb{R}^{4 \times 1}$ which is not expressible as $av_1 + bv_2 + cv_3 + dv_4$ for any $a, b, c, d \in \mathbb{R}$.
   [Hint: Compute the RREF of $A = [v_1^t \ v_2^t \ v_3^t \ v_4^t]$.]

1.6 Determinant

There are two important quantities associated with a square matrix. One is the trace, and the other is the determinant.

The sum of all diagonal entries of a square matrix is called the **trace** of the matrix. That is, if $A = [a_{ij}] \in \mathbb{F}^{m \times m}$, then

$$\text{tr}(A) = a_{11} + \cdots + a_{nn} = \sum_{k=1}^{n} a_{kk}.$$

In addition to $\text{tr}(I_m) = m$ and $\text{tr}(0) = 0$, the trace satisfies the following properties:

1. $\text{tr}(\alpha A) = \alpha \, \text{tr}(A)$ for each $\alpha \in \mathbb{F}$.
2. $\text{tr}(A^t) = \text{tr}(A)$ and $\text{tr}(A^*) = \overline{\text{tr}(A)}$.
3. $\text{tr}(A + B) = \text{tr}(A) + \text{tr}(B)$.
4. $\text{tr}(AB) = \text{tr}(BA)$.
5. $\text{tr}(A^*A) = 0$ iff $\text{tr}(AA^*) = 0$ iff $A = 0$.

Observe that (4) does not assert that the trace of a product is equal to the product of their traces. Further, $\text{tr}(A^*A) = \sum_{i=1}^{m} \sum_{j=1}^{m} |a_{ij}|^2$ proves (5).

The **determinant** of a square matrix $A = [a_{ij}] \in \mathbb{F}^{n \times n}$, written as $\det(A)$, is defined inductively as follows:

1. If $n = 1$, then $\det(A) = a_{11}$.
2. If $n > 1$, then $\det(A) = \sum_{j=1}^{n} (-1)^{1+j} a_{1j} \det(A_{1j})$,
   where $A_{1j} \in \mathbb{F}^{(n-1) \times (n-1)}$ is obtained from A by deleting the first row and the jth column of A.

When $A = [a_{ij}]$ is written showing all its entries, we also write $\det(A)$ by replacing the two big closing brackets [and] by two vertical bars | and |. For a 2×2 matrix, its determinant is seen as follows:

$$\begin{vmatrix} a_{11} & a_{12} \\ a_{21} & a_{22} \end{vmatrix} = (-1)^{1+1} a_{11} \det[a_{22}] + (-1)^{1+2} a_{12} \det[a_{21}] = a_{11}a_{22} - a_{12}a_{21}.$$

Similarly, for a 3×3 matrix, we need to compute three 2×2 determinants. For instance,

$$\begin{vmatrix} 1 & 2 & 3 \\ 2 & 3 & 1 \\ 3 & 1 & 2 \end{vmatrix} = (-1)^{1+1} \times 1 \times \begin{vmatrix} 3 & 1 \\ 1 & 2 \end{vmatrix} + (-1)^{1+2} \times 2 \times \begin{vmatrix} 2 & 1 \\ 3 & 2 \end{vmatrix} + (-1)^{1+3} \times 3 \times \begin{vmatrix} 2 & 3 \\ 3 & 1 \end{vmatrix}$$

$$= 1 \times \begin{vmatrix} 3 & 1 \\ 1 & 2 \end{vmatrix} - 2 \times \begin{vmatrix} 2 & 1 \\ 3 & 2 \end{vmatrix} + 3 \times \begin{vmatrix} 2 & 3 \\ 3 & 1 \end{vmatrix}$$

$$= (3 \times 2 - 1 \times 1) - 2 \times (2 \times 2 - 1 \times 3) + 3 \times (2 \times 1 - 3 \times 3)$$

$$= 5 - 2 \times 1 + 3 \times (-7) = -18.$$

For a lower triangular matrix, we see that

$$\begin{vmatrix} a_{11} & & & & \\ a_{12} & a_{22} & & & \\ a_{13} & a_{23} & a_{33} & & \\ \vdots & & & \ddots & \\ a_{n1} & \cdots & & & a_{nn} \end{vmatrix} = a_{11} \begin{vmatrix} a_{22} & & & \\ a_{23} & a_{33} & & \\ \vdots & & \ddots & \\ a_{n1} & \cdots & & a_{nn} \end{vmatrix} = \cdots = a_{11}a_{22} \cdots a_{nn}.$$

In general, the determinant of any triangular matrix (upper or lower) is the product of its diagonal entries. In particular, the determinant of a diagonal matrix is also the product of its diagonal entries. Thus, if I is the identity matrix of order n, then $\det(I) = 1$ and $\det(-I) = (-1)^n$.

Our definition of determinant *expands* the determinant in the first row. In fact, the same result may be obtained by expanding it in any other row or even in any column. To mention some similar properties of the determinant, we introduce some terminology.

Let $A \in \mathbb{F}^{n \times n}$. The sub-matrix of A obtained by deleting the ith row and the jth column is called the (i, j)th **minor** of A and is denoted by A_{ij}.

The (i, j)th **co-factor** of A is $(-1)^{i+j}\det(A_{ij})$; it is denoted by $C_{ij}(A)$. Sometimes, when the matrix A is fixed in a context, we write $C_{ij}(A)$ as C_{ij}.

The **adjugate** of A is the $n \times n$ matrix obtained by taking transpose of the matrix whose (i, j)th entry is $C_{ij}(A)$; it is denoted by $\mathrm{adj}(A)$. That is, $\mathrm{adj}(A) \in \mathbb{F}^{n \times n}$ is the matrix whose (i, j)th entry is the (j, i)th co-factor $C_{ji}(A)$.

Also, we write $A_i(x)$ for the matrix obtained from A by replacing its ith row by a row vector x of appropriate size.

Let $A \in \mathbb{F}^{n \times n}$. Let $i, j, k \in \{1, \ldots, n\}$. Let $E[i, j]$, $E_\alpha[i]$ and $E_\alpha[i, j]$ be the elementary matrices of order n for $1 \le i \ne j \le n$ and a nonzero scalar α. Then, the following statements are true.

1. $\det(E[i, j]\, A) = -\det(A)$.
2. $\det(E_\alpha[i]\, A) = \alpha \det(A)$.
3. $\det(E_\alpha[i, j]\, A) = \det(A)$.
4. If some row of A is the zero vector, then $\det(A) = 0$.
5. If one row of A is a scalar multiple of another row, then $\det(A) = 0$.
6. For any $i \in \{1, \ldots, n\}$, $\det(A_i(x + y)) = \det(A_i(x)) + \det(A_i(y))$.
7. $\det(A^t) = \det(A)$ and $\det(A^*) = \overline{\det(A)}$.
8. If A is a triangular matrix, then $\det(A)$ is equal to the product of the diagonal entries of A.
9. $\det(AB) = \det(A)\det(B)$ for any matrix $B \in \mathbb{F}^{n \times n}$.
10. $A\,\mathrm{adj}(A) = \mathrm{adj}(A)A = \det(A)\, I$.
11. A is invertible iff $\det(A) \ne 0$.

Elementary column operations are operations similar to row operations, but with columns instead of rows. Notice that since $\det(A^t) = \det(A)$, the facts concerning elementary row operations also hold true if elementary column operations are used. Using elementary (row and column) operations, the *computational complexity* for evaluating a determinant can be reduced drastically. The trick is to bring a matrix to a triangular form by using elementary row operations, so that the determinant of the ensuing triangular matrix can be computed easily.

Example 1.5

$$\begin{vmatrix} 1 & 0 & 0 & 1 \\ -1 & 1 & 0 & 1 \\ -1 & -1 & 1 & 1 \\ -1 & -1 & -1 & 1 \end{vmatrix} \overset{R1}{=} \begin{vmatrix} 1 & 0 & 0 & 1 \\ 0 & 1 & 0 & 2 \\ 0 & -1 & 1 & 2 \\ 0 & -1 & -1 & 2 \end{vmatrix} \overset{R2}{=} \begin{vmatrix} 1 & 0 & 0 & 1 \\ 0 & 1 & 0 & 2 \\ 0 & 0 & 1 & 4 \\ 0 & 0 & -1 & 4 \end{vmatrix} \overset{R3}{=} \begin{vmatrix} 1 & 0 & 0 & 1 \\ 0 & 1 & 0 & 2 \\ 0 & 0 & 1 & 4 \\ 0 & 0 & 0 & 8 \end{vmatrix} = 8.$$

Here, $R1 = E_1[2, 1]$, $E_1[3, 1]$, $E_1[4, 1]$; $R2 = E_1[3, 2]$, $E_1[4, 2]$; $R3 = E_1[4, 3]$. □

Example 1.6 See that the following is true, for verifying Property (6) as mentioned above:

$$\begin{vmatrix} 3 & 1 & 2 & 4 \\ -1 & 1 & 0 & 1 \\ -1 & -1 & 1 & 1 \\ -1 & -1 & -1 & 1 \end{vmatrix} = \begin{vmatrix} 1 & 0 & 0 & 1 \\ -1 & 1 & 0 & 1 \\ -1 & -1 & 1 & 1 \\ -1 & -1 & -1 & 1 \end{vmatrix} + \begin{vmatrix} 2 & 1 & 2 & 3 \\ -1 & 1 & 0 & 1 \\ -1 & -1 & 1 & 1 \\ -1 & -1 & -1 & 1 \end{vmatrix}.$$

□

Exercises for Sect. 1.6

1. Prove the properties of the trace of a matrix as mentioned in the text.
2. For each $n > 2$, construct an $n \times n$ nonzero matrix where no row is a scalar multiple of another row but its determinant is 0.
3. Is it true that if $\det(A) = 0$, then among the rows $R_1, \ldots, R_n$ of A, some row, say, R_i can be written as $R_i = \alpha_1 R_1 + \cdots + \alpha_{i-1} R_{i-1} + \alpha_{i+1} R_{i+1} + \cdots + \alpha_n R_n$ for some scalars α_j? What about columns instead of rows?
4. Let $a_1, \ldots, a_n \in \mathbb{C}$. Let A be the $n \times n$ matrix whose first row has all entries as 1 and whose kth row has entries $a_1^{k-1}, \ldots, a_n^{k-1}$ in that order. Show that $\det(A) = (-1)^{n(n-1)/2} \Pi_{1 \le i < j \le n}(a_i - a_j)$.
5. Compute A^{-1} using $\mathrm{adj}(A)$, where $A = \begin{bmatrix} 1 & 0 & 0 & 1 \\ -1 & 1 & 0 & 1 \\ -1 & -1 & 1 & 1 \\ -1 & -1 & -1 & 1 \end{bmatrix}$.
6. Let $A, B \in \mathbb{F}^{3 \times 3}$ with $\det(A) = 2$ and $\det(B) = 3$. Determine
   (a) $\det(4A)$ (b) $\det(AB)$ (c) $\det(5AB)$ (d) $\det(2A^{-1}B)$
7. Let $A, B \in \mathbb{F}^{2 \times 2}$, and let $E = [e_{ij}]$ with $e_{11} = e_{22} = 0$, $e_{12} = e_{21} = 1$. Show that if $B = EA$, then $\det(A + B) = \det(A) + \det(B)$.
8. Give examples of A and B in $\mathbb{R}^{2 \times 2}$ so that $\det(A + B) \ne \det(A) + \det(B)$.

1.7 Computing Inverse of a Matrix

The adjugate property of the determinant provides a way to compute the inverse of a matrix, provided it is invertible. However, it is very inefficient. We may use elementary row operations to compute the inverse. Our computation of the inverse bases on the following fact.

Theorem 1.2 *A square matrix is invertible iff it is a product of elementary matrices.*

Proof Since elementary matrices are invertible, so is their product. Conversely, suppose that A is an invertible matrix. Let EA^{-1} be the RREF of A^{-1}. If EA^{-1} has a zero row, then $E = EA^{-1}A$ also has a zero row. But E is a product of elementary matrices, which is invertible; it does not have a zero row. Therefore, EA^{-1} does not have a zero row. Then, each row in the square matrix EA^{-1} has a pivot. But the only square matrix in RREF having a pivot at each row is the identity matrix. Therefore, $EA^{-1} = I$. That is, $A = E$, a product of elementary matrices. ∎

Now, suppose a matrix A is invertible. Then, there exist elementary matrices $E_1, \ldots, E_m$ such that $A = E_1 \cdots E_m$. It follows that $A^{-1} = E_m^{-1} \cdots E_1^{-1} I$ and $E_m^{-1} \cdots E_1^{-1} A = I$. Since I is in RREF, it follows from the uniqueness of RREF that the row reduced form of A is I. Therefore, if a sequence of elementary row operations applied on A results in I, then the same sequence applied on I will result in A^{-1}.

This suggests applying the same elementary row operations on both A and I simultaneously so that A is reduced to its RREF. For this purpose, we introduce the notion of an augmented matrix. If $A \in \mathbb{F}^{m \times n}$ and $B \in \mathbb{F}^{m \times k}$, then the matrix $[A \mid B] \in \mathbb{F}^{m \times (n+k)}$ obtained from A and B by writing first all the columns of A and then the columns of B, in that order, is called an **augmented matrix**. The vertical bar shows the separation of columns of A and of B, though conceptually unnecessary.

For computing the inverse of a matrix, we start with the augmented matrix $[A \mid I]$. We then apply elementary row operations for reducing A to its RREF, while simultaneously applying the same operations on the entries of I. This means we pre-multiply the matrix $[A \mid I]$ with a product E of elementary matrices so that EA is in RREF. In block form, our result is the augmented matrix $[EA \mid EI]$. If A is invertible, then $EA = I$ and then $EI = A^{-1}$. Once the A portion has been reduced to I, the I portion is A^{-1}. On the other hand, if A is not invertible, then its RREF will have at least one zero row.

Example 1.7 Determine inverses of the following matrices, if possible:

$$
A = \begin{bmatrix} 1 & -1 & 2 & 0 \\ -1 & 0 & 0 & 2 \\ 2 & 1 & -1 & -2 \\ 1 & -2 & 4 & 2 \end{bmatrix}, \quad
B = \begin{bmatrix} 1 & -1 & 2 & 0 \\ -1 & 0 & 0 & 2 \\ 2 & 1 & -1 & -2 \\ 0 & -2 & 0 & 2 \end{bmatrix}.
$$

We augment A with an identity matrix to get

$$
\left[\begin{array}{cccc|cccc}
1 & -1 & 2 & 0 & 1 & 0 & 0 & 0 \\
-1 & 0 & 0 & 2 & 0 & 1 & 0 & 0 \\
2 & 1 & -1 & -2 & 0 & 0 & 1 & 0 \\
1 & -2 & 4 & 2 & 0 & 0 & 0 & 1
\end{array}\right]
$$

Next, we use elementary row operations in order to reduce A to its RREF. Since $a_{11} = 1$, it is a pivot. To zero-out the other entries in the first column, we use the sequence of elementary row operations $E_1[2, 1]$, $E_{-2}[3, 1]$, $E_{-1}[4, 1]$, and obtain

$$
\left[\begin{array}{cccc|cccc}
\boxed{1} & -1 & 2 & 0 & 1 & 0 & 0 & 0 \\
0 & -1 & 2 & 2 & 1 & 1 & 0 & 0 \\
0 & 3 & -5 & -2 & -2 & 0 & 1 & 0 \\
0 & -1 & 2 & 2 & -1 & 0 & 0 & 1
\end{array}\right]
$$

The pivot is -1 in $(2, 2)$ position. Use $E_{-1}[2]$ to make the pivot 1. And then, use $E_1[1, 2]$, $E_{-3}[3, 2]$, $E_1[4, 2]$ to zero-out all non-pivot entries in the pivotal column:

$$
\left[\begin{array}{cccc|cccc}
\boxed{1} & 0 & 0 & -2 & 0 & -1 & 0 & 0 \\
0 & \boxed{1} & -2 & -2 & -1 & -1 & 0 & 0 \\
0 & 0 & \boxed{1} & 4 & 1 & 3 & 1 & 0 \\
0 & 0 & 0 & 0 & -2 & -1 & 0 & 1
\end{array}\right]
$$

Since a zero row has appeared in the A portion of the augmented matrix, we conclude that A is not invertible. You see that the second portion of the augmented matrix has no meaning now. However, it records the elementary row operations which were carried out in the reduction process. Verify that this matrix is equal to

$$E_1[4, 2]\, E_{-3}[3, 2]\, E_1[1, 2]\, E_{-1}[2]\, E_{-1}[4, 1]\, E_{-2}[3, 1]\, E_1[2, 1].$$

and that the first portion is equal to this matrix times A.

For B, we proceed similarly. The augmented matrix $[B \mid I]$ with the first pivot looks like:

$$\begin{bmatrix} \boxed{1} & -1 & 2 & 0 & 1 & 0 & 0 & 0 \\ -1 & 0 & 0 & 2 & 0 & 1 & 0 & 0 \\ 2 & 1 & -1 & -2 & 0 & 0 & 1 & 0 \\ 0 & -2 & 0 & 2 & 0 & 0 & 0 & 1 \end{bmatrix}$$

The sequence of elementary row operations $E_1[2, 1]$, $E_{-2}[3, 1]$ yields

$$\begin{bmatrix} \boxed{1} & -1 & 2 & 0 & 1 & 0 & 0 & 0 \\ 0 & -1 & 2 & 2 & 1 & 1 & 0 & 0 \\ 0 & 3 & -5 & -2 & -2 & 0 & 1 & 0 \\ 0 & -2 & 0 & 2 & 0 & 0 & 0 & 1 \end{bmatrix}$$

Next, the pivot is -1 in $(2, 2)$ position. Use $E_{-1}[2]$ to get the pivot as 1. And then, $E_1[1, 2]$, $E_{-3}[3, 2]$, $E_2[4, 2]$ gives

$$\begin{bmatrix} \boxed{1} & 0 & 0 & -2 & 0 & -1 & 0 & 0 \\ 0 & \boxed{1} & -2 & -2 & -1 & -1 & 0 & 0 \\ 0 & 0 & 1 & 4 & 1 & 3 & 1 & 0 \\ 0 & 0 & -4 & -2 & -2 & -2 & 0 & 1 \end{bmatrix}$$

Next pivot is 1 in $(3, 3)$ position. Now, $E_2[2, 3]$, $E_4[4, 3]$ produces

$$\begin{bmatrix} \boxed{1} & 0 & 0 & -2 & 0 & -1 & 0 & 0 \\ 0 & \boxed{1} & 0 & 6 & 1 & 5 & 2 & 0 \\ 0 & 0 & \boxed{1} & 4 & 1 & 3 & 1 & 0 \\ 0 & 0 & 0 & 14 & 2 & 10 & 4 & 1 \end{bmatrix}$$

Next pivot is 14 in $(4, 4)$ position. Use $E_{1/14}[4]$ to get the pivot as 1. Use $E_2[1, 4]$, $E_{-6}[2, 4]$, $E_{-4}[3, 4]$ to zero-out the entries in the pivotal column:

$$\begin{bmatrix} \boxed{1} & 0 & 0 & 0 & 2/7 & 3/7 & 4/7 & 1/7 \\ 0 & \boxed{1} & 0 & 0 & 1/7 & 5/7 & 2/7 & -3/7 \\ 0 & 0 & \boxed{1} & 0 & 3/7 & 1/7 & -1/7 & -2/7 \\ 0 & 0 & 0 & \boxed{1} & 1/7 & 5/7 & 2/7 & 1/14 \end{bmatrix}$$

$$\text{Thus, } B^{-1} = \tfrac{1}{7} \begin{bmatrix} 2 & 3 & 4 & 1 \\ 1 & 5 & 2 & -3 \\ 3 & 1 & -1 & -2 \\ 1 & 5 & 2 & 1/2 \end{bmatrix}. \text{ Verify that } B^{-1}B = BB^{-1} = I. \qquad \square$$

Observe that if a matrix is not invertible, then our algorithm for reduction to RREF produces a pivot in the I portion of the augmented matrix.

Exercises for Sect. 1.7

1. Compute the inverses of the following matrices, if possible:

(a) $\begin{bmatrix} 2 & 1 & 2 \\ 1 & 3 & 1 \\ -1 & 1 & 2 \end{bmatrix}$ (b) $\begin{bmatrix} 1 & 4 & -6 \\ -1 & -1 & 3 \\ 1 & -2 & 3 \end{bmatrix}$ (c) $\begin{bmatrix} 3 & 1 & 1 & 2 \\ 1 & 2 & 0 & 1 \\ 1 & 1 & 2 & -1 \\ -2 & 1 & -1 & 3 \end{bmatrix}$

2. Let $A = \begin{bmatrix} 2 & 1 \\ 6 & 4 \end{bmatrix}$. Express A and A^{-1} as products of elementary matrices.

3. Given matrices $A = \begin{bmatrix} 5 & 2 \\ 3 & 1 \end{bmatrix}$ and $B = \begin{bmatrix} 3 & 4 \\ 1 & 2 \end{bmatrix}$, find matrices X and Y such that $AX = B$ and $YA = B$. [Hint: Both A and B are invertible.]

4. Let $A = \begin{bmatrix} 0 & 1 & 0 \\ 0 & 0 & 1 \\ 1 & -b & -c \end{bmatrix}$, where $b, c \in \mathbb{C}$. Show that $A^{-1} = bI + cA$.

5. Show that if A is an upper triangular invertible matrix, then so is A^{-1}.
6. Show that if A is a lower triangular invertible matrix, then so is A^{-1}.
7. Can every square matrix be written as a sum of two invertible matrices?
8. Can every invertible matrix be written as a sum of two non-invertible matrices?

1.8 Problems

1. Let A and D be matrices of order n. Suppose that D is a diagonal matrix. Describe the products AD and DA.
2. Give examples of matrices A and B so that
   (a) $A^2 - B^2 \neq (A + B)(A - B)$ (b) $(A + B)^2 \neq A^2 + 2AB + B^2$
3. Find nonzero matrices A, B and C such that $A \neq B$ and $AC = BC$.
4. Construct a 2×2 matrix A with each entry nonzero so that $A^2 = 0$. Can such a matrix A satisfy $A^* = A$?
5. Construct a matrix A and column vectors $x \neq y$ such that $Ax = Ay$. Can such a matrix A be invertible?
6. Construct matrices A and B so that $AB = A$ and $B \neq I$. Is such a matrix A invertible?
7. Give infinitely many 2×2 invertible matrices A such that $A^{-1} = A^t$.
8. Give infinitely many 2×2 matrices A satisfying $A^2 = I$.

9. Let $A \in \mathbb{F}^{n \times n}$ satisfy $A^2 = 0$. Show that $I - A$ is invertible. What is the inverse of $I - A$?

10. Let $A \in \mathbb{F}^{n \times n}$ satisfy $A^{k+1} = 0$ for some $k \in \mathbb{N}$. Show that $I - A$ is invertible. [Hint: What could be the inverse?]

11. For $n \in \mathbb{N}$, compute A^n, where $A = [a_{ij}] \in \mathbb{F}^{2 \times 2}$ is given by
    (a) $a_{11} = a_{12} = a_{21} = a_{22} = 1$ (b) $a_{11} = -a_{12} = -a_{21} = a_{22} = \frac{1}{2}$

12. A square matrix A is called *idempotent* iff $A^2 = A$. Show the following:

    (a) Let A be a 3×3 matrix whose third row has each entry as $\frac{1}{2}$ and all other entries as $\frac{1}{4}$. Then, A is idempotent.
    (b) A diagonal matrix with diagonal entries in $\{0, 1\}$ is idempotent.
    (c) If A is idempotent and B is invertible, then $B^{-1}AB$ is idempotent.
    (d) If A is idempotent, then so is $I - A$.
    (e) If A is idempotent, then $I + A$ is invertible.
    (f) If A is a square matrix with $A^2 = I$, then $B = \frac{1}{2}(I + A)$ and $C = \frac{1}{2}(I - A)$ are idempotent, and $BC = 0$.

13. Let B be the RREF of an $m \times n$ matrix A of rank r. Show the following:

    (a) Let $u_1, \ldots, u_n$ be the columns of B. Let $k_1, \ldots, k_r$ be the column indices of all pivotal columns in B. If $v = \alpha_1 u_1 + \cdots + \alpha_n u_n$ for some scalars α_s, then $v = \beta_1 u_{k_1} + \cdots + \beta_r u_{k_r}$ for some scalars β_s. Moreover, among these r columns $u_{k_1}, \ldots, u_{k_r}$, no vector u_{k_i} can be expressed as $u_{k_i} = \gamma_1 u_{k_1} + \cdots + \gamma_{i-1} u_{k_{i-1}} + \gamma_{i+1} u_{k_{i+1}} + \cdots + \gamma_r u_{k_r}$ for scalars γ_s.
    (b) Let $w_{j_1}, \ldots, w_{j_r}$ be the rows of A which have become the pivotal rows in B. Then, any row w of A can be written as $w = \alpha_1 w_{j_1} + \cdots + \alpha_r w_{j_r}$ for some scalars α_s. Further, no vector w_{j_i} can be written as $w_{j_i} = \beta_1 w_{j_1} + \cdots + \beta_{i-1} w_{j_{i-1}} + \beta_{i+1} w_{j_{i+1}} + \cdots + \beta_r w_{j_r}$ for scalars β_s.

14. Let $v_1, \ldots, v_n \in \mathbb{F}^{n \times 1}$ be such that the matrix $P = [v_1 \ \cdots \ v_n]$ is invertible. Let $v \in \mathbb{F}^{n \times 1}$. For $\alpha_1, \ldots, \alpha_n \in \mathbb{F}$, show that $v = \alpha_1 v_1 + \cdots + \alpha_n v_n$ iff $P^{-1}v = [\alpha_1, \ldots, \alpha_n]^t$. Connect this result to the RREF of the augmented matrix $[v_1 \ \cdots \ v_n \mid v]$.

15. Let $A = \begin{bmatrix} 2 & 1 & 1 \\ 4 & 1 & 3 \\ 6 & 4 & 5 \end{bmatrix}$. Determine a product E of elementary matrices so that $U = EA$ is upper triangular. With $L = E^{-1}$, we have $A = LU$. What type of matrix is L?

16. Let $A \in \mathbb{C}^{n \times n}$. Show the following:
    (a) If $\operatorname{tr}(AA^*) = 0$, then $A = 0$.   (b) If $A^*A = A^2$, then $A^* = A$.

17. Suppose each entry of an $n \times n$ matrix A is an integer. If $\det(A) = \pm 1$, then prove that each entry of A^{-1} is also an integer.

18. The *Vandermonde matrix* with numbers $a_1, \ldots, a_{n+1}$ is a matrix of order $n + 1$ whose ith row is $[a_1^{i-1}, \ldots, a_{n+1}^{i-1}]$ for $1 \le i \le n + 1$. Show that the determinant of such a matrix is given by the product $\Pi_{i<j}(a_i - a_j)$.

19. Let $A, E \in \mathbb{F}^{m \times m}, B, F \in \mathbb{F}^{m \times n}, C, G \in \mathbb{F}^{n \times m}$, and let $D, H \in \mathbb{F}^{n \times n}$. Show that

$$\begin{bmatrix} A & B \\ C & D \end{bmatrix} \begin{bmatrix} E & F \\ G & H \end{bmatrix} = \begin{bmatrix} AE + BG & AF + BH \\ CE + DG & CF + DH \end{bmatrix}.$$

20. Let A, B, C, D, E, F, G, H be as in Problem 19. Show that

$$\begin{bmatrix} A & B \\ C & D \end{bmatrix}^{-1} = \begin{bmatrix} (A - BD^{-1}C)^{-1} & A^{-1}B(CA^{-1}B - D)^{-1} \\ (CA^{-1}B - D)^{-1}CA^{-1} & (D - CA^{-1}B)^{-1} \end{bmatrix}$$

provided all the inverses exist.

21. Let A, B, C, D, E, F, G, H be as in Problem 19. Let $M = \begin{bmatrix} A & B \\ C & D \end{bmatrix}$. Show the following:

  (a) If $B = 0$ or $C = 0$, then $\det(M) = \det(A) \cdot \det(B)$.
  (b) If $\det(A) \neq 0$, then $\det(M) = \det(A) \cdot \big(\det(D - CA^{-1}B)\big)$.
    [Hint: Consider the matrix $M \begin{bmatrix} I_m & -A^{-1}B \\ 0 & I_n \end{bmatrix}$.]

22. Elementary column operations work with the columns of a matrix in a similar fashion as elementary row operations work with the rows. Let A be an $m \times n$ matrix. Show the following:

  (a) $A\,E[i, j]$ is obtained from A by exchanging the ith and jth columns.
  (b) $A\,E_\alpha[i]$ is obtained from A by multiplying α with the ith column.
  (c) $A\,E_\alpha[i, j]$ is obtained from A by adding α times ith column to the jth column.

23. What simpler form can you obtain by using a sequence of elementary column operations on a matrix which is already in RREF?

24. Let E_{ij} denote the $n \times n$ matrix with its (i, j)th entry as 1, and all other entries as 0. Let A be any $n \times n$ matrix. Show the following:

  (a) If $i \neq j$, then $E_{ij}^2 = 0$ and $(I + E_{ij})^{-1} = I - E_{ij}$.
  (b) $E_{ii}^2 = E_{ii}$ and $(I + E_{ii})^{-1} = I - \frac{1}{2}E_{ii}$.
  (c) If $AE_{ij} = E_{ij}A$ for all $i, j \in \{1, \ldots, n\}$, then $a_{11} = a_{22} = \cdots = a_{nn}$, and $a_{ij} = 0$ for $i \neq j$.
  (d) If $AB = BA$ for all invertible $n \times n$ matrices B, then $A = \alpha I$ for some scalar α.

25. Let $A = [a_{ij}] \in \mathbb{F}^{n \times n}, B = \mathrm{diag}(1, 2, \ldots, n), C = \mathrm{diag}(a_{11}, a_{22}, \ldots, a_{nn})$, and let $D \in \mathbb{F}^{n \times n}$ have its first row as $[1, 2, \ldots, n]$, all diagonal entries as 1, and all other entries as 0. Show the following

  (a) If $AB = BA$, then $a_{ij} = 0$ for all $i \neq j$.
  (b) If $CD = DC$, then $a_{11} = a_{22} = \cdots = a_{nn}$.
  (c) If $AM = MA$ for all invertible matrices $M \in \mathbb{F}^{n \times n}$, then $A = \alpha I$ for some scalar α.

Chapter 2
Systems of Linear Equations

2.1 Linear Independence

In the reduction to RREF, why some rows are reduced to zero rows and why the others are not reduced to zero rows? Similarly, in the RREF of a matrix, why some columns are pivotal and others are not?

Recall that a row vector in $\mathbb{F}^{1 \times n}$ and a column vector in $\mathbb{F}^{n \times 1}$ are both written uniformly as an n-tuple $(a_1, \ldots, a_n)$ in $\mathbb{F}^n$. Such an n-tuple of scalars from $\mathbb{F}$ is interpreted as either a row vector with n components or a column vector with n components, as the case demands. Thus, an n-tuple of scalars is called a *vector* in $\mathbb{F}^n$.

The sum of two vectors from $\mathbb{F}^n$ and the multiplication of a vector from $\mathbb{F}^n$ by a scalar follow those of the row and/or column vectors. That is, for $\beta \in \mathbb{F}, (a_1, \ldots, a_n), (b_1, \ldots, b_n) \in \mathbb{F}^n$, we define

$$(a_1, \ldots, a_n) + (b_1, \ldots, b_n) = (a_1 + b_1, \ldots, a_n + b_n),$$
$$\beta(a_1, \ldots, a_n) = (\beta a_1, \ldots, \beta a_n).$$

Let $v_1, \ldots, v_m \in \mathbb{F}^n$. The vector $\alpha_1 v_1 + \cdots + \alpha_m v_m \in \mathbb{F}^n$ is called a **linear combination** of $v_1, \ldots, v_m$, where $\alpha_1, \ldots, \alpha_m$ are some scalars from $\mathbb{F}$.

By taking some particular scalars α_i, if the sum $\alpha_1 v_1 + \cdots + \alpha_m v_m$ evaluates to a vector v, then we also say that v is a linear combination of the vectors $v_1, \ldots, v_m$.

For example, one linear combination of $v_1 = [1, 1]$ and $v_2 = [1, -1]$ is

$$2[1, 1] + 1[1, -1].$$

This linear combination evaluates to $[3, 1]$. We say that $[3, 1]$ is a linear combination of v_1 and v_2. Is $[4, -2]$ a linear combination of v_1 and v_2? Yes, since

$$[4, -1] = 1[1, 1] + 3[1, -1].$$

© The Author(s), under exclusive license to Springer Nature Switzerland AG 2021
A. Singh, *Introduction to Matrix Theory*,
https://doi.org/10.1007/978-3-030-80481-7_2

In fact, every vector in $\mathbb{F}^{1\times 2}$ is a linear combination of v_1 and v_2. Reason:

$$[a, b] = \tfrac{a+b}{2}[1, 1] + \tfrac{a-b}{2}[1, -1].$$

However, every vector in $\mathbb{F}^{1\times 2}$ is not a linear combination of $[1, 1]$ and $[2, 2]$. Reason? Any linear combination of these two vectors is a scalar multiple of $[1, 1]$. Then, $[1, 0]$ is not a linear combination of these two vectors.

Let $A \in \mathbb{F}^{m\times n}$ be a matrix of rank r. Let $u_{i_1}, \ldots, u_{i_r}$ be the r columns of A that correspond to the pivotal columns in its RREF. Then among these r columns, no u_{i_j} is a linear combination of others, and each other column in A is a linear combination of these r columns.

Similarly, let $w_{k_1}, \ldots, w_{k_r}$ be the r rows in A that correspond to the r nonzero rows in its RREF (monitoring the row exchanges). Then among these r rows, no w_{k_j} is a linear combination of others, and each other row of A is a linear combination of these r rows.

The vectors $v_1, \ldots, v_m$ in $\mathbb{F}^n$ are called **linearly dependent** iff at least one of them is a linear combination of others. The vectors are called **linearly independent** iff none of them is a linear combination of others.

For example, $[1, 1], [1, -1], [3, 1]$ are linearly dependent vectors since $[3, 1] = 2[1, 1] + [1, -1]$, whereas $[1, 1], [1, -1]$ are linearly independent vectors in $\mathbb{F}^{1\times 2}$.

If $\alpha_1 = \cdots = \alpha_m = 0$, then the linear combination $\alpha_1 v_1 + \cdots + \alpha_m v_m$ evaluates to 0. That is, the zero vector can always be written as a *trivial* linear combination. However, a non-trivial linear combination of some vectors can evaluate to 0. For instance, $2[1, 1] + [1, -1] - [3, 1] = 0$. We guess that this can happen for linearly dependent vectors, but may not happen for linearly independent vectors.

Suppose the vectors $v_1, \ldots, v_m$ are linearly dependent. Then, one of them, say, v_i is a linear combination of others. That is, for some scalars α_j,

$$v_i = \alpha_1 v_1 + \cdots + \alpha_{i-1} v_{i-1} + \alpha_{i+1} v_{i+1} + \cdots + \alpha_m v_m.$$

Then,

$$\alpha_1 v_1 + \cdots + \alpha_{i-1} v_{i-1} + (-1)v_i + \alpha_{i+1} v_{i+1} + \cdots + \alpha_m v_m = 0.$$

Here, we see that a linear combination becomes zero, where at least one of the coefficients, that is, the ith one, is nonzero. That is, a non-trivial linear combination of $v_1, \ldots, v_m$ exists which evaluates to 0.

Conversely, suppose that we have scalars $\beta_1, \ldots, \beta_m$ not all zero such that

$$\beta_1 v_1 + \cdots + \beta_m v_m = 0.$$

Say, the kth scalar β_k is nonzero. Then,

$$v_k = -\frac{1}{\beta_k}\big(\beta_1 v_1 + \cdots + \beta_{k-1} v_{k-1} + \beta_{k+1} v_{k+1} + \cdots + \beta_m v_m\big).$$

That is, the vectors $v_1, \ldots, v_m$ are linearly dependent.

Thus, we have proved the following:

$v_1, \ldots, v_m$ are linearly dependent

iff $\alpha_1 v_1 + \cdots + \alpha_m v_m = 0$ for scalars $\alpha_1, \ldots, \alpha_m$ not all zero

iff the zero vector can be written as a *non-trivial linear combination* of $v_1, \ldots, v_m$.

The same may be expressed in terms of linear independence.

Theorem 2.1 *Vectors* $v_1, \ldots, v_m \in \mathbb{F}^n$ *are linearly independent iff for all* $\alpha_1, \ldots, \alpha_m \in \mathbb{F}$,

$$\alpha_1 v_1 + \cdots \alpha_m v_m = 0 \ \text{implies that} \ \alpha_1 = \cdots = \alpha_m = 0.$$

Theorem 2.1 provides a way to determine whether a finite number of vectors are linearly independent or not. You start with a linear combination of the given vectors and equate it to 0. Then, use the laws of addition and scalar multiplication to derive that each coefficient in that linear combination is 0. Once you succeed, you conclude that the given vectors are linearly independent. On the other hand, if it is not possible to derive that each coefficient is 0, then from the proof of this impossibility you will be able to express one of the vectors as a linear combination of the others. And this would prove that the given vectors are linearly dependent.

Example 2.1 Are the vectors $[1, 1, 1], [2, 1, 1], [3, 1, 0]$ linearly independent?

We start with an arbitrary linear combination and equate it to the zero vector. Then, we solve the resulting linear equations to determine whether all the coefficients are necessarily 0 or not. So, let

$$a\,[1, 1, 1] + b\,[2, 1, 1] + c\,[3, 1, 0] = [0, 0, 0].$$

Comparing the components, we have

$$a + 2b + 3c = 0, a + b + c = 0, a + b = 0.$$

The last two equations imply that $c = 0$. Substituting in the first, we see that $a + 2b = 0$. This and the equation $a + b = 0$ give $b = 0$. Then, it follows that $a = 0$. We conclude that the given vectors are linearly independent. □

Example 2.2 Are the vectors $[1, 1, 1], [2, 1, 1], [3, 2, 2]$ linearly independent?

Clearly, the third one is the sum of the first two. So, the given vectors are linearly dependent.

To illustrate our method, we start with an arbitrary linear combination and equate it to the zero vector. We then solve the resulting linear equations to determine whether all the coefficients are necessarily 0 or not. As earlier, let

$$a\,[1, 1, 1] + b\,[2, 1, 1] + c\,[3, 2, 2] = [0, 0, 0].$$

Comparing the components, we have

$$a + 2b + 3c = 0, a + b + 2c = 0, a + b + 2c = 0.$$

The last equation is redundant. From the first and the second, we have

$$b + c = 0.$$

We may choose $b = 1, c = -1$ to satisfy this equation. Then from the second equation, we have $a = 1$. Our starting equation says that the third vector is the sum of the first two.   □

Be careful with the direction of implication here. Your workout must be in the following form:

Assume $\alpha_1 v_1 + \cdots + \alpha_m v_m = 0$. Then $\cdots$ Then $\alpha_1 = \cdots = \alpha_m = 0$.
And that would prove linear independence.

To see how linear independence is helpful, consider the following system of linear equations:

$$\begin{array}{rrrl} x_1 & +2x_2 & -3x_3 & = 2 \\ 2x_1 & -x_2 & +2x_3 & = 3 \\ 4x_1 & +3x_2 & -4x_3 & = 7 \end{array}$$

Here, we find that the third equation is redundant, since 2 times the first plus the second gives the third. That is, the third one linearly depends on the first two. You can of course choose any other equation here as linearly depending on the other two, but that is not important and that may not be always possible. Now, take the row vectors of coefficients of the unknowns along with the right-hand side, as in the following:

$$v_1 = [1, 2, -3, 2], \quad v_2 = [2, -1, 2, 3], \quad v_3 = [4, 3, -4, 7].$$

We see that $v_3 = 2v_1 + v_2$, as it should be. That is, the vectors v_1, v_2, v_3 are linearly dependent. But the vectors v_1, v_2 are linearly independent. Thus, solving the given system of linear equations is the same thing as solving the system with only first two equations.

For solving linear systems, it is of primary importance to find out which equations linearly depend on others. Once determined, such equations can be thrown away, and the rest can be solved.

As to our opening questions, now we know that in the RREF of a matrix, a column that corresponds to a non-pivotal column is a linear combination of the columns that correspond to the pivotal columns. Similarly, a zero row corresponds to one (monitoring row exchanges) which is a linear combination of the nonzero (pivotal) rows in the RREF. We will see this formally in the next section.

Exercises for Sect. 2.1

1. Check whether the given vectors are linearly independent:

   (a) $(1, 2, 6), (-1, 3, 4), (-1, -4, 2)$ in $\mathbb{R}^3$
   (b) $(1, 0, 2, 1), (1, 3, 2, 1), (4, 1, 2, 2)$ in $\mathbb{C}^4$

2. Suppose that u, v, w are linearly independent in $\mathbb{C}^5$. Are the following vectors linearly independent?

   (a) $u + v, v + w, w + u$
   (b) $u - v, v - w, w - u$
   (c) $u, v + \alpha w, w$, where α is a nonzero real number.
   (d) $u + v, v + 2w, u + \alpha v, v - \alpha w$, where α is a complex number.

3. Give three linearly dependent vectors in $\mathbb{R}^2$ such that none of the three is a scalar multiple of another.
4. Suppose S is a finite set of vectors, and some $v \in S$ is not a linear combination of other vectors in S. Is S linearly independent?
5. Prove that nonzero vectors $v_1, \ldots, v_m \in \mathbb{F}^n$ are linearly dependent iff there exists $k > 1$ such that the vectors $v_1, \ldots, v_{k-1}$ are linearly independent and v_k is a linear combination of $v_1, \ldots, v_{k-1}$.

2.2 Determining Linear Independence

If a vector w is a linear combination of vectors v_1, v_2 and if v_1, v_2 are linear combinations of u_1, u_2, then w is a linear combination of u_1, u_2. We look at the reduction to RREF via this principle.

The reduction to RREF is achieved by performing elementary row operations. The row exchanges do not disturb linear dependence or independence except reordering the row vectors. So, let us look at the other types of elementary row operations.

Suppose the kth row of an $m \times n$ matrix A is a linear combination of its first r rows. To fix notation, let $v_1, \ldots, v_m$ be the rows of A. We have scalars $\alpha_1, \ldots, \alpha_r$ and $k > r$ such that

$$v_k = \alpha_1 v_1 + \cdots + \alpha_r v_r.$$

Now, the rows of $E_\beta[i] A$ are $v_1, \ldots, v_{i-1}, \beta v_i, v_{i+1}, \ldots, v_n$ for some $\beta \neq 0$.
If $1 \le i \le r$, then $v_k = \alpha_1 v_1 + \cdots \alpha_{i-1} v_{i-1} + \frac{\alpha_i}{\beta_i} \beta_i v_i + \alpha_{i+1} v_{i+1} + \cdots + \alpha_r v_r$.
If $i = k$, then $\beta v_k = \alpha_1 \beta v_1 + \cdots + \alpha_r \beta v_r$.
If $i > r, i \neq k$, then $v_k = \alpha_1 v_1 + \cdots + \alpha_r v_r$.
In any case, the kth row of $E_\beta[i] A$ is a linear combination of the first r rows of $E_\beta[i] A$.

Similarly, the rows of $E_\beta[i, j] A$ are $v_1, \ldots, v_{i-1}, v_i + \beta v_j, v_{i+1}, \ldots, v_m$. Suppose that $1 \le j \le r$.
If $1 \le i \le r$, then $v_k = \alpha_1 v_1 + \cdots + \alpha_i (v_i + \beta v_j) + \cdots \alpha_r v_r - \alpha_i \beta v_j$.
If $i = k > r$, then $v_k + \beta v_j = \alpha_1 v_1 + \cdots + \alpha_i (v_i + \beta v_j) + \cdots \alpha_r v_r$.
If $i > r, i \neq k$, then $v_k = \alpha_1 v_1 + \cdots + \alpha_r v_r$.
In any case, if $1 \le j \le r$, then the kth row of $E_\beta[i, j] A$ is a linear combination of the first r of its rows. We summarize these facts as follows.

Observation 2.1 Let $A \in \mathbb{F}^{m \times n}$. Let $v_1, \ldots, v_m$ be the rows of A in that order. Let $E \in \mathbb{F}^{m \times m}$ be a product of elementary matrices of the types $E_\beta[i]$ and/or $E_\beta[i, j]$, where $1 \leq i, j \leq r$. Let $w_1, \ldots, w_m$ be the rows of EA in that order. Let $k > r$. If the kth row v_k of A is a linear combination of the vectors $v_1, \ldots, v_r$, then the kth row w_k in EA is also a linear combination of $w_1, \ldots, w_r$.

In the notation of Observation 2.1, you can also show that if any row vector v is a linear combination of $v_1, \ldots, v_r$, then it is a linear combination of $w_1, \ldots, w_r$ and vice versa.

Theorem 2.2 *Let* $A \in \mathbb{F}^{m \times n}$ *be a matrix of rank* r. *Let* v_i *denote the* ith *row of* A. *Suppose that the rows* $v_{i_1}, \ldots, v_{i_r}$ *have become the pivotal rows* $w_1, \ldots, w_r$, *respectively, in the RREF of* A. *Let* v *be any row of* A *other than* $v_{i_1}, \ldots, v_{i_r}$. *Then, the following are true:*

(1) The vectors $v_{i_1}, \ldots, v_{i_r}$ *are linearly independent.*
(2) The row v *of* A *has become a zero row in the RREF of* A.
(3) The row v *is a linear combination of* $v_{i_1}, \ldots, v_{i_r}$.
(4) The row v *is a linear combination of* $w_1, \ldots, w_r$.

Proof Monitoring the row exchanges, it can be found out which rows have become zero rows and which rows have become the pivotal rows. Assume, without loss of generality, that no row exchanges have been performed during reduction of A to its RREF B. Then, $B = EA$, where E is a product of elementary matrices of the forms $E_\beta[i]$ or $E_\beta[i, j]$. The first r rows in B are the pivotal rows, i.e. $i_1 = 1, i_2 = 2, \ldots, i_r = r$. So, suppose that $v_1, \ldots, v_r$ have become the pivotal rows $w_1, \ldots, w_r$ in B, respectively.
(1) If one of $v_1, \ldots, v_r$, say, v_k is a linear combination of the others, then by Observation 2.1, the pivotal row w_k is a linear combination of other pivotal rows. But this is not possible. Hence among the vectors $v_1, \ldots, v_r$, none of them is a linear combination of the others. Therefore, $v_1, \ldots, v_r$ are linearly independent.
(2) Since rank$(A) = r$ and $v_1, \ldots, v_r$ have become the pivotal rows, no other row is a pivotal row. That is, all other rows of A, including v, have become zero rows.
(3) The vectors $w_{r+1}, \ldots, w_m$ are the zero rows in B. Each of them is a linear combination of the pivotal rows $w_1, \ldots, w_r$. Now, the vectors $w_1, \ldots, w_r$ are rows in B, and $v_1, \ldots, v_r$ are the corresponding rows in $A = E^{-1}B$, where E^{-1} is a product of elementary matrices of the forms $E_\beta[i]$ or $E_\beta[i, j]$ with $1 \leq j \leq r$. By Observation 2.1, each of $v_{r+1}, \ldots, v_m$ is a linear combination of $v_1, \ldots, v_r$.
(4) During row reduction, elementary row operations use the pivotal rows. Therefore, each of the vectors $v_1, \ldots, v_r$ is a linear combination of $w_1, \ldots, w_r$; and each of the vectors $w_1, \ldots, w_r$ is a linear combination of $v_1, \ldots, v_r$. Then, it follows from (3) that v_{r+k} is a linear combination of $w_1, \ldots, w_r$ also. ∎

Theorem 2.2 can be used to determine linear independence of vectors. Given vectors $v_1, \ldots, v_m \in \mathbb{F}^{1 \times n}$, we form a matrix A by taking these as its rows. During the reduction of A to its RREF, if a zero row appears, then the vectors are linearly dependent. Else, the rank of A turns out to be m; consequently, the vectors are linearly independent.

Example 2.3 To determine whether the vectors $[1, 1, 0, 1]$, $[0, 1, 1, -1]$, and $[1, 3, 2, -1]$ are linearly independent or not, we form a matrix with the given vectors as its rows and then reduce it to its RREF. It is as follows.

$$\begin{bmatrix} \boxed{1} & 1 & 0 & 1 \\ 0 & 1 & 1 & -1 \\ 1 & 3 & 2 & -1 \end{bmatrix} \xrightarrow{E_{-1}[3,1]} \begin{bmatrix} \boxed{1} & 1 & 0 & 1 \\ 0 & \boxed{1} & 1 & -1 \\ 0 & 2 & 2 & -2 \end{bmatrix} \xrightarrow{R1} \begin{bmatrix} \boxed{1} & 0 & -1 & 2 \\ 0 & \boxed{1} & 1 & -1 \\ 0 & 0 & 0 & -4 \end{bmatrix}$$

$$\xrightarrow{R2} \begin{bmatrix} \boxed{1} & 0 & -1 & 0 \\ 0 & \boxed{1} & 1 & 0 \\ 0 & 0 & 0 & \boxed{1} \end{bmatrix}.$$

Here, $R1 = E_{-1}[1, 2], E_{-2}[3, 2]$ and $R2 = E_{-1/4}[3], E_{-2}[1, 3], E_{1}[2, 3]$.

The last matrix is in RREF in which there is no zero row; each row has a pivot. So, the original vectors are linearly independent. $\qquad\square$

Example 2.4 Are the vectors $[1, 1, 0, 1]$, $[0, 1, 1, -1]$ and $[2, -1, -3, 5]$ linearly independent?

We construct a matrix with the vectors as rows and reduce it to RREF.

$$\begin{bmatrix} \boxed{1} & 1 & 0 & 1 \\ 0 & 1 & 1 & -1 \\ 2 & -1 & -3 & 5 \end{bmatrix} \xrightarrow{E_{-2}[3,1]} \begin{bmatrix} \boxed{1} & 1 & 0 & 1 \\ 0 & \boxed{1} & 1 & -1 \\ 0 & -3 & -3 & 3 \end{bmatrix} \xrightarrow{R1} \begin{bmatrix} \boxed{1} & 0 & -1 & -2 \\ 0 & \boxed{1} & 1 & -1 \\ 0 & 0 & 0 & 0 \end{bmatrix}.$$

Here, $R1 = E_{-1}[1, 2], E_{3}[3, 2]$. Since a zero row has appeared, the original vectors are linearly dependent. Also, notice that no row exchanges were carried out in the reduction process. So, the third vector is a linear combination of the first two, which are linearly independent. $\qquad\square$

Does reduction to RREF change the linear dependence or linear independence of columns of a matrix?

Theorem 2.3 *Let* $E \in \mathbb{F}^{m \times m}$ *be invertible. Let* $v_1, \ldots, v_k, v \in \mathbb{F}^{m \times 1}$ *and* $\alpha_1, \ldots, \alpha_k \in \mathbb{F}$.

(1) $v = \alpha_1 v_1 + \cdots \alpha_k v_k$ *iff* $Ev = \alpha_1 Ev_1 + \cdots + \alpha_k Ev_k$.
(2) $v_1, \ldots, v_k$ *are linearly independent iff* $Ev_1, \ldots, Ev_k$ *are linearly independent.*

Proof (1) If $v = \alpha_1 v_1 + \cdots + \alpha_k v_k$, then multiplying E on the left, we have $Ev = \alpha_1 Ev_1 + \cdots + \alpha_k Ev_k$. Conversely, if $Ev = \alpha_1 Ev_1 + \cdots + \alpha_k Ev_k$, then multiplying E^{-1} on the left we have $v = \alpha_1 v_1 + \cdots + \alpha_k v_k$.
(2) Vectors $v_1, \ldots, v_k$ are linearly dependent iff there exist scalars $\beta_1, \ldots, \beta_k$ not all zero such that $\beta_1 v_1 + \cdots \beta_k v_k = 0$. By (1), this happens iff there exist scalars $\beta_1, \ldots, \beta_k$ not all zero such that $\beta_1 Ev_1 + \cdots \beta_k Ev_k = 0$ iff the vectors $Ev_1, \ldots, Ev_k$ are linearly dependent. $\qquad\blacksquare$

Since a product of elementary matrices is invertible, Theorem 2.3 implies that reduction to RREF does not change the linear dependence or linear independence of column vectors of a matrix. This, along with Observation 1.2, yields the following result.

Theorem 2.4 *Let* $v_1, \ldots, v_n \in \mathbb{F}^{m \times 1}$. *Let* $A = [v_1 \cdots v_n] \in \mathbb{F}^{m \times n}$ *be the matrix whose jth column is* v_j. *Then, the following are true:*

(1) Vectors $v_1, \ldots, v_n$ *are linearly independent iff* $\text{rank}(A) = n$ *iff each column is a pivotal column in the RREF of* A.

(2) If the kth column in the RREF of A *is a non-pivotal column, then it has the form* $[a_1, a_2, \ldots, a_j, 0, \ldots 0]^t$ *where* j *is the number of pivotal columns to the left of this kth column,* $j < k$, *and* $v_k = a_1 v_1 + \cdots + a_j v_j$.

Given vectors $v_1, \ldots, v_n \in \mathbb{F}^{m \times 1}$, we form the matrix $A = [v_1 \cdots v_n]$ and then apply Theorem 2.4. If the vectors are linearly dependent, then it will help us in finding out a subset of vectors which are linearly independent. It will also let us know how the dependent ones could be expressed as linear combinations of the independent ones. If the vectors are linearly independent, then all columns of A in its RREF will turn out to be pivotal columns. The advantage in working with columns is that the RREF shows explicitly how a linearly dependent vector depends on the linearly independent ones.

Moreover, if instead of column vectors, we are given with row vectors, then we may work with their transposes. We solve Example 2.4 once more to illustrate this point.

Example 2.5 To determine whether $u_1 = [1, 1, 0, 1]$, $u_2 = [0, 1, 1, -1]$ $u_3 = [2, -1, -3, 5]$ are linearly independent or not, we form the matrix $\begin{bmatrix} u_1^t & u_2^t & u_3^t \end{bmatrix}$ and then reduce it to its RREF, as in the following:

$$\begin{bmatrix} 1 & 0 & 2 \\ 1 & 1 & -1 \\ 0 & 1 & -3 \\ 1 & -1 & 5 \end{bmatrix} \xrightarrow{R1} \begin{bmatrix} \boxed{1} & 0 & 2 \\ 0 & 1 & -3 \\ 0 & 1 & -3 \\ 0 & -1 & 3 \end{bmatrix} \xrightarrow{R2} \begin{bmatrix} \boxed{1} & 0 & 2 \\ 0 & \boxed{1} & -3 \\ 0 & 0 & 0 \\ 0 & 0 & 0 \end{bmatrix}.$$

Here, $R1 = E_{-1}[2, 1]$, $E_{-1}[4, 1]$ and $R2 = E_{-1}[3, 2] E_1[4, 2]$. The first and second columns are pivotal, and the third is non-pivotal. The components of the third column in the RREF show that $u_3^t = 2 u_1^t - 3 u_2^t$. Thus, $u_3 = 2 u_1 - 3 u_2$. $\square$

We also use the phrases such as linear dependence or independence for sets of vectors. Given a set of vectors $A = \{v_1, \ldots, v_m\}$, we say that A is a **linearly independent set** iff the vectors $v_1, \ldots, v_m$ are linearly independent. Our method of determining linear independence gives rise to the following useful result.

Theorem 2.5 *Any set containing more than n vectors from* $\mathbb{F}^n$ *is linearly dependent.*

Proof Without loss of generality, let $S \subseteq \mathbb{F}^{n \times 1}$ have more than n vectors. So, let $v_1, \ldots, v_{n+1}$ be distinct vectors from S. Consider the (augmented) matrix

$$A = [v_1 \quad \cdots \quad v_n \mid v_{n+1}].$$

Let B be the RREF of A. The reduction to RREF shows that there can be at most n pivots in B and the last column in B is a non-pivotal column. By Observation 1.3, v_{n+1} is a linear combination of the pivotal columns, which are among $v_1, \ldots, v_n$. Therefore, A is linearly dependent. ∎

Exercises for Sect. 2.2

1. Using elementary row operations, determine whether the given vectors are linearly dependent or independent in each of the following cases, by taking the given vectors as rows of a matrix.

   (a) $[1, 2, 3], [4, 5, 6], [7, 8, 9]$
   (b) $[1, 0, -1, 2, -3], [-2, 1, 2, 4, -1], [3, 0, -1, 1, 1], [2, 1, 1, -1, -2]$
   (c) $[1, 0, -1, 2, -3], [-2, 1, 2, 4, -1], [3, 0, -1, 1, 1], [2, -1, 0, -7, 3]$
   (d) $[1, i, -1, 1 - i], [i, -1, -i, 1 + i], [2, 0, 1, i], [1 + i, 1 - i, -1, -i]$

2. Solve Exercise 1 by forming a matrix with its columns as the transposes of the given vectors. If the vectors turn out to be linearly dependent, then express one of them as a linear combination of the others.

3. Let $A = [u_1, u_2, u_3, u_4, u_5] \in \mathbb{F}^{4 \times 5}$. In each of the following cases, determine the RREF of A.

   (a) u_1, u_2, u_3 are linearly independent; $u_4 = u_1 + 2u_2, u_5 = u_1 - u_2 + 2u_3$.
   (b) u_1, u_2, u_4 are linearly independent; $u_3 = u_1 + 2u_2, u_5 = u_1 - u_2 + 2u_4$.
   (c) u_1, u_3, u_5 are linearly independent; $u_2 = u_1 + 2u_3, u_4 = u_1 - u_3 + u_5$.

4. Answer the following questions with justification:

   (a) Is every subset of a linearly independent set linearly independent?
   (b) Is every subset of a linearly dependent set linearly dependent?
   (c) Is every superset of a linearly independent set linearly independent?
   (d) Is every superset of a linearly dependent set linearly dependent?
   (e) Is union of two linearly independent sets linearly independent?
   (f) Is union of two linearly dependent sets linearly dependent?
   (g) Is intersection of two linearly independent sets linearly independent?
   (h) Is intersection of two linearly dependent sets linearly dependent?

2.3 Rank of a Matrix

Recall that the rank of a matrix A, denoted by $\mathrm{rank}(A)$, is the number of pivots in the RREF of A. If $\mathrm{rank}(A) = r$, then there are r number of linearly independent columns in A and other columns are linear combinations of these r columns. The linearly independent columns correspond to the pivotal columns. Also, there exist r number of linearly independent rows of A such that other rows are linear combinations of these r rows. The linearly independent rows correspond to the pivotal rows, assuming that we have monitored the row exchanges during the reduction of A to its RREF.

It raises a question. Suppose for a matrix A, we find r number of linearly independent rows such that other rows are linear combinations of these r rows. Can it happen that there are also k rows which are linearly independent and other rows are linear combinations of these k rows, and that $k \neq r$?

Theorem 2.6 *Let $A \in \mathbb{F}^{m \times n}$. There exists a unique number r with $0 \leq r \leq \min\{m, n\}$ such that A has r number of linearly independent rows, and other rows of A are linear combinations of these r ones. Moreover, such a number r is equal to the rank of A.*

Proof Let $r = \text{rank}(A)$. Theorem 2.2 implies that there exist r rows of A which are linearly independent and the other rows are linear combinations of these r rows.

Conversely, suppose there exist r rows of A which are linearly independent and the other rows are linear combinations of these r rows. So, let $i_1, \ldots, i_r$ be the indices of these r numbers of linearly independent rows of A. Consider the matrix

$$B = E[1, i_1] \, E[2, i_2] \, \cdots \, E[r, i_r] \, A.$$

In the matrix B, those r linearly independent rows of A have become the first r rows, and other rows are now placed as $(r + 1)$th row onwards. In the RREF of B, the first r rows are pivotal rows, and other rows are zero rows. The matrix B has been obtained from A by elementary row operations. By the uniqueness of RREF (Theorem 1.1), A also has the same RREF as does B. The number of pivots in this RREF is r. Therefore, $\text{rank}(A) = r$.

Moreover, when $A = 0$, the zero matrix, the number of pivots in A is 0. And if A is a nonzero matrix, then in the RREF of A, there exists at least one pivot. The number of pivots cannot exceed the number of rows; also, it cannot exceed the number of columns. Therefore, $r = \text{rank}(A)$ is a number between 0 and $\min\{m, n\}$. ∎

Now if there are k number of linearly independent columns in an $m \times n$ matrix A, where the other columns are linear combinations of these k ones, then the same thing happens about rows in A^t. By Theorem 2.6, $\text{rank}(A^t)$ is equal to this k.

Conventionally, the maximum number of linearly independent rows of a matrix is called the **row rank** of a matrix. Similarly, the maximum number of linearly independent columns of a matrix is called its **column rank**. As we see, due to Theorem 2.6, the row rank and the column rank of a matrix are well defined. To connect the row rank and the column rank, we prove the following result.

Theorem 2.7 *Let $A \in \mathbb{F}^{m \times n}$. Then, $\text{rank}(A^t) = \text{rank}(A)$. Consequently, both the row rank of A and the column rank of A are equal to $\text{rank}(A)$.*

Proof Let $B = EA^t$ be the RREF of A^t, where E is a suitable product of elementary matrices. Let $\text{rank}(A^t) = r$. Then, there are r number of pivots in B. The pivotal columns in B are $e_1, \ldots, e_r \in \mathbb{F}^{n \times 1}$. Since E is invertible, by Theorem 2.3, $E^{-1}e_1, \ldots, E^{-1}e_r$ are linearly independent columns of A^t, and the other columns of A^t are linear combinations of these r ones. Then, $(E^{-1}e_1)^t, \ldots, (E^{-1}e_r)^t$ are the r number of linearly independent rows of A and other rows are linear combinations of these r rows. By Theorem 2.6, $\text{rank}(A) = r$. This proves that if $\text{rank}(A^t) = r$, then $\text{rank}(A) = r$.

Conversely, suppose $\text{rank}(A) = r$. As $(A^t)^t = A$, we have $\text{rank}((A^t)^t) = r$. By what we have just proved, $\text{rank}(A^t) = r$.

Therefore, $\text{rank}(A^t) = r$ iff $\text{rank}(A) = r$. That is, $\text{rank}(A^t) = \text{rank}(A)$.

Then, it follows from Theorem 2.6 that the row rank of A is equal to $\text{rank}(A)$, which is equal to $\text{rank}(A^t)$, and that is equal to the column rank of A. ∎

Example 2.6 Let $A = \begin{bmatrix} 1 & 1 & 1 & 2 & 1 \\ 1 & 2 & 1 & 1 & 1 \\ 3 & 5 & 3 & 4 & 3 \\ -1 & 0 & -1 & -3 & -1 \end{bmatrix}$. We compute its RREF as follows:

$$\begin{bmatrix} \boxed{1} & 1 & 1 & 2 & 1 \\ 1 & 2 & 1 & 1 & 1 \\ 3 & 5 & 3 & 4 & 3 \\ -1 & 0 & -1 & -3 & -1 \end{bmatrix} \xrightarrow{R1} \begin{bmatrix} \boxed{1} & 1 & 1 & 2 & 1 \\ 0 & 1 & 0 & -1 & 0 \\ 0 & 2 & 0 & -2 & 0 \\ 0 & 1 & 0 & -1 & 0 \end{bmatrix} \xrightarrow{R2} \begin{bmatrix} \boxed{1} & 0 & 1 & 3 & 1 \\ 0 & \boxed{1} & 0 & -1 & 0 \\ 0 & 0 & 0 & 0 & 0 \\ 0 & 0 & 0 & 0 & 0 \end{bmatrix}.$$

Here, $R1 = E_{-1}[2, 1]$, $E_{-3}[3, 1]$, $E_1[4, 1]$; $R2 = E_{-1}[1, 2]$, $E_{-2}[3, 2]$, $E_{-1}[4, 2]$.

Thus, $\text{rank}(A) = 2$.

Also, we see that the first two rows of A are linearly independent, and

$$row(3) = row(1) + 2 \times row(2), \quad row(4) = row(2) - 2 \times row(1).$$

Thus, the row rank of A is 2.

From the RREF of A, we observe that the first two columns of A are linearly independent, and

$$col(3) = col(1), \quad col(4) = 3 \times col(1) - 2 \times col(2), \quad col(4) = col(1).$$

Therefore, the column rank of A is also 2. □

Example 2.7 Determine the rank of the matrix A in Example 1.4, and point out which rows of A are linear combinations of other rows and which columns are linear combinations of other columns, by reducing A to its RREF.

From Example 1.4, we have seen that

$$A = \begin{bmatrix} \boxed{1} & 1 & 2 & 0 \\ 3 & 5 & 7 & 1 \\ 1 & 5 & 4 & 5 \\ 2 & 8 & 7 & 9 \end{bmatrix} \xrightarrow{E} \begin{bmatrix} \boxed{1} & 0 & 3/2 & 0 \\ 0 & \boxed{1} & 1/2 & 0 \\ 0 & 0 & 0 & \boxed{1} \\ 0 & 0 & 0 & 0 \end{bmatrix}.$$

The row operation E is given by

$$E = E_{-3}[2, 1],\ E_{-1}[3, 1],\ E_{-2}[4, 1]\, E_{-1}[2, 1],\ E_{-4}[3, 2],\ E_{-6}[4, 2],$$
$$E_{1/2}[1, 3],\ E_{-1/2}[2, 3],\ E_{-6}[4, 3].$$

We see that $\mathrm{rank}(A) = 3$, the number of pivots in the RREF of A. In this reduction, no row exchanges have been used. Thus, the first three rows of A are the required linearly independent rows. The fourth row is a linear combination of these three rows. In fact,

$$row(4) = 3\,row(1) + (-1)\,row(2) + 2\,row(3).$$

The RREF also says that the third column is a linear combination of first and second. Notice that the coefficients in such a linear combination are given by the entries of the third column in the RREF. As we have seen earlier,

$$col(3) = \tfrac{3}{2}col(1) + \tfrac{1}{2}col(2). \hspace{3cm} \square$$

Since each entry of A^* is the complex conjugate of the corresponding entry of A^t, Theorem 2.7 implies that

$$\mathrm{rank}(A^*) = \mathrm{rank}(A^t) = \mathrm{rank}(A).$$

Theorem 2.3 also implies that the column rank is well defined and is equal to the rank. We generalize this theorem a bit.

Theorem 2.8 *Let $A \in \mathbb{F}^{m \times n}$. Let $P \in \mathbb{F}^{m \times m}$ and $Q \in \mathbb{F}^{n \times n}$ be invertible matrices. Then,*

$$\mathrm{rank}(PAQ) = \mathrm{rank}(PA) = \mathrm{rank}(AQ) = \mathrm{rank}(A).$$

Proof Theorem 2.3 implies that the column rank of PAQ is same as the column rank of AQ. Therefore, $\mathrm{rank}(PAQ) = \mathrm{rank}(AQ)$. Also, since Q^t is invertible, we have $\mathrm{rank}(AQ) = \mathrm{rank}(Q^t A^t) = \mathrm{rank}(A^t) = \mathrm{rank}(A)$. ∎

In general, when the matrix product PAQ is well defined, we have

$$\mathrm{rank}(PAQ) \leq \mathrm{rank}(A)$$

irrespective of whether P and Q are invertible or not.

Exercises for Sect. 2.3

1. Determine rank r of the following matrices. Find out the r linearly independent rows and also the r linearly independent columns of the matrix. And then, express $4 - r$ rows as linear combinations of those r rows and $5 - r$ columns as linear combinations of those r columns.

(a) $\begin{bmatrix} 1 & 2 & 1 & 1 & 1 \\ 3 & 5 & 3 & 4 & 3 \\ 1 & 1 & 1 & 2 & 1 \\ 5 & 8 & 5 & 7 & 5 \end{bmatrix}$
(b) $\begin{bmatrix} 0 & 2 & 1 & 0 & 1 \\ 3 & 0 & 3 & 0 & 3 \\ 1 & 1 & 0 & 2 & 0 \\ 5 & 0 & 5 & 0 & 5 \end{bmatrix}$
(c) $\begin{bmatrix} 1 & 0 & 0 & 1 & 0 \\ 0 & 5 & 0 & 4 & 0 \\ 0 & 0 & 1 & 0 & 1 \\ 0 & 8 & 0 & 7 & 0 \end{bmatrix}$

2. Let $A \in \mathbb{F}^{n \times n}$. Prove that A is invertible iff $\text{rank}(A) = n$ iff $\det(A) \neq 0$.
3. Let $A \in \mathbb{F}^{m \times n}$. Let $P \in \mathbb{F}^{m \times m}$. When does the RREF of PA coincide with the RREF of A?
4. Let $u, v \in \mathbb{F}^{n \times 1}$. What is the rank of the matrix uv^t?
5. Show that if $A \in \mathbb{F}^{n \times n}$ has rank 1, then there exist $u, v \in \mathbb{F}^{n \times 1}$ such that $A = uv^t$.

2.4 Solvability of Linear Equations

We now use matrices to settle some issues regarding solvability of systems of linear equations, also called linear systems. A **linear system** with m equations in n unknowns looks as follows:

$$a_{11}x_1 + a_{12}x_2 + \cdots a_{1n}x_n = b_1$$
$$a_{21}x_1 + a_{22}x_2 + \cdots a_{2n}x_n = b_2$$
$$\vdots$$
$$a_{m1}x_1 + a_{m2}x_2 + \cdots a_{mn}x_n = b_m$$

Solving such a linear system amounts to determining the unknowns $x_1, \ldots, x_n$ with known scalars a_{ij} and b_i. Using the abbreviation $x = [x_1, \ldots, x_n]^t$, $b = [b_1, \ldots, b_m]^t$ and $A = [a_{ij}]$, the system can be written as

$$Ax = b.$$

Here, $A \in \mathbb{F}^{m \times n}$, $b \in \mathbb{F}^{m \times 1}$, and x is an unknown vector in $\mathbb{F}^{n \times 1}$. We also say that the matrix A is the **system matrix** of the linear system $Ax = b$.

There is a slight deviation from our accepted symbolism. In case of linear systems, we write b as a column vector and x_i are unknown scalars.

Notice that if the system matrix $A \in \mathbb{F}^{m \times n}$, then the linear system $Ax = b$ has m number of equations and n number of unknowns.

A **solution** of the system $Ax = b$ is any vector $y \in \mathbb{F}^{n \times 1}$ such that $Ay = b$. In such a case, if $y = [c_1, \ldots, c_n]^t$, then c_i is called as the **value of the unknown** x_i in the solution y. A solution of the system is also written informally as

$$x_1 = c_1, \ldots, x_n = c_n.$$

If $y = [c_1, \ldots, c_n]^t$ is a solution of $Ax = b$, and $v_1, \ldots, v_n \in \mathbb{F}^{m \times 1}$ are the columns of A, then from the above system, we see that

$$c_1 v_1 + \cdots + c_n v_n = b.$$

Conversely, if b can be written this way for some scalars $c_1, \ldots, c_n$, then the vector $y = [c_1, \ldots, c_n]^t$ is a solution of $Ax = b$. So, we conclude that

$Ax = b$ has a solution iff b can be written as a linear combination of the columns of A.

Corresponding to the linear system, $Ax = b$ is the **homogeneous system**

$$Ax = 0.$$

The homogeneous system always has a solution since $y = 0$ is a solution. If y is a solution of $Ax = 0$, then so is αy for any scalar α. Therefore, the homogeneous system has infinitely many solutions when it has a nonzero solution.

Theorem 2.9 *Let $A \in \mathbb{F}^{m \times n}$, and let $b \in \mathbb{F}^{m \times 1}$. Then, the following statements are true:*

(1) If $[A' \mid b']$ is obtained from $[A \mid b]$ by applying a finite sequence of elementary row operations, then each solution of $Ax = b$ is a solution of $A'x = b'$ and vice versa.

(2) **(Consistency)** *$Ax = b$ has a solution iff $\operatorname{rank}([A \mid b]) = \operatorname{rank}(A)$.*

(3) If u is a (particular) solution of $Ax = b$, then each solution of $Ax = b$ is given by $u + y$, where y is a solution of the homogeneous system $Ax = 0$.

(4) If $r = \operatorname{rank}([A \mid b]) = \operatorname{rank}(A) < n$, then there are $n - r$ unknowns which can take arbitrary values and other r unknowns can be determined from the values of these $n - r$ unknowns.

(5) If $m < n$, then the homogeneous system has infinitely many solutions.

(6) $Ax = b$ has a unique solution iff $\operatorname{rank}([A \mid b]) = \operatorname{rank}(A) = n$.

(7) If $m = n$, then $Ax = b$ has a unique solution iff $\det(A) \neq 0$.

(8) **(Cramer's Rule)** *If $m = n$ and $\det(A) \neq 0$, then the solution of $Ax = b$ is given by $x_j = \det(A_j(b))/\det(A)$ for each $j \in \{1, \ldots, n\}$.*

Proof (1) If $[A' \mid b']$ has been obtained from $[A \mid b]$ by a finite sequence of elementary row operations, then $A' = EA$ and $b' = Eb$, where E is the product of corresponding elementary matrices. The matrix E is invertible. Now, $A'x = b'$ iff $EAx = Eb$ iff $Ax = E^{-1}Eb = b$.

(2) Due to (1), we assume that $[A \mid b]$ is in RREF. Suppose $Ax = b$ has a solution. If there is a zero row in A, then the corresponding entry in b is also 0. Therefore, there is no pivot in b. Hence, $\operatorname{rank}([A \mid b]) = \operatorname{rank}(A)$.

Conversely, suppose that $\operatorname{rank}([A \mid b]) = \operatorname{rank}(A) = r$. Then, there is no pivot in b. That is, b is a non-pivotal column in $[A \mid b]$. Thus, b is a linear combination of pivotal columns, which are some columns of A. Therefore, $Ax = b$ has a solution.

(3) Let u be a solution of $Ax = b$. Then, $Au = b$. Now, z is a solution of $Ax = b$ iff $Az = b$ iff $Az = Au$ iff $A(z - u) = 0$ iff $z - u$ is a solution of $Ax = 0$. That is, each solution z of $Ax = b$ is expressed in the form $z = u + y$ for a solution y of the homogeneous system $Ax = 0$.

(4) Let $\operatorname{rank}([A \mid b]) = \operatorname{rank}(A) = r < n$. By (2), there exists a solution. Due to (3), we consider solving the corresponding homogeneous system. Due to (1), assume that A is in RREF. There are r number of pivots in A and $m - r$ number of zero rows. Omit all the zero rows; it does not affect the solutions. Write the system as

linear equations. Rewrite the equations by keeping the unknowns corresponding to pivots on the left-hand side and taking every other term to the right-hand side. The unknowns corresponding to pivots are now expressed in terms of the other $n - r$ unknowns. For obtaining a solution, we may arbitrarily assign any values to these $n - r$ unknowns, and the unknowns corresponding to the pivots get evaluated by the equations.

(5) Let $m < n$. Then, $r = \text{rank}(A) \leq m < n$. Consider the homogeneous system $Ax = 0$. By (4), there are $n - r \geq 1$ number of unknowns which can take arbitrary values, and other r unknowns are determined accordingly. Each such assignment of values to the $n - r$ unknowns gives rise to a distinct solution resulting in infinite number of solutions of $Ax = 0$.

We may generate infinite number of solutions as follows. Since $r < n$, there exists an equation where on the left-hand side is an unknown corresponding to a pivot, and on the right-hand side there is an unknown that does not correspond to a pivot having a nonzero coefficient. Fix one such equation. Now, assign this unknown a nonzero value and all other unknowns zero. By varying the assigned values, we get infinite number of values for the unknown on the left-hand side.

(6) It follows from (3) and (4). The unique solution is given by $x = A^{-1}b$.

(7) If $A \in \mathbb{F}^{n \times n}$, then it is invertible iff $\text{rank}(A) = n$ iff $\det(A) \neq 0$. Then, the statement follows from (6).

(8) Recall that $A_j(b)$ is the matrix obtained from A by replacing the jth column of A with the vector b. Since $\det(A) \neq 0$, by (6), $Ax = b$ has a unique solution, say $y \in \mathbb{F}^{n \times 1}$. Write the identity $Ay = b$ in the form.

$$y_1 \begin{bmatrix} a_{11} \\ \vdots \\ a_{n1} \end{bmatrix} + \cdots + y_j \begin{bmatrix} a_{1j} \\ \vdots \\ a_{nj} \end{bmatrix} + \cdots + y_n \begin{bmatrix} a_{1n} \\ \vdots \\ a_{nn} \end{bmatrix} = \begin{bmatrix} b_1 \\ \vdots \\ b_n \end{bmatrix}.$$

This gives

$$y_1 \begin{bmatrix} a_{11} \\ \vdots \\ a_{n1} \end{bmatrix} + \cdots + \begin{bmatrix} (y_j a_{1j} - b_1) \\ \vdots \\ (y_j a_{nj} - b_n) \end{bmatrix} + \cdots + y_n \begin{bmatrix} a_{1n} \\ \cdots \\ a_{nn} \end{bmatrix} = 0.$$

In this sum, the jth vector is a linear combination of other vectors, where $-y_j$s are the coefficients. Therefore,

$$\begin{vmatrix} a_{11} & \cdots & (y_j a_{1j} - b_1) & \cdots & a_{1n} \\ & & \vdots & & \\ a_{n1} & \cdots & (y_j a_{nj} - b_n) & \cdots & a_{nn} \end{vmatrix} = 0.$$

From Property (6) of the determinant, it follows that

$$y_j \begin{vmatrix} a_{11} & \cdots & a_{1j} & \cdots & a_{1n} \\ & & \vdots & & \\ a_{n1} & \cdots & a_{nj} & \cdots & a_{nn} \end{vmatrix} - \begin{vmatrix} a_{11} & \cdots & b_1 & \cdots & a_{1n} \\ & & \vdots & & \\ a_{n1} & \cdots & b_n & \cdots & a_{nn} \end{vmatrix} = 0.$$

Therefore, $y_j = \det(A_j(b))/\det(A)$.   ∎

A linear system $Ax = b$ is said to be **consistent** iff $\operatorname{rank}([A \mid b]) = \operatorname{rank}(A)$. Theorem 2.9 says that only consistent systems have solutions. The homogeneous system
$Ax = 0$ is always consistent. It always has a solution, namely the zero solution,
$x = 0$, which is also called the **trivial solution**.

For a matrix $A \in \mathbb{F}^{m \times n}$, the number $n - \operatorname{rank}(A)$ is called the **nullity** of A and is
denoted by $\operatorname{null}(A)$.

The unknowns that correspond to the pivots are called the **basic variables**, and
the other unknowns are called the **free variables**. Thus, there are $\operatorname{rank}(A)$ number of
basic variables and $\operatorname{null}(A)$ number of free variables, which may be assigned arbitrary
values for obtaining a solution. The statement in Theorem 2.9(4) is informally stated
as follows:

A consistent system has $\operatorname{null}(A)$ *number of linearly independent solutions.*

This terminology of *linearly independent solutions* is meaningful when we consider a solution of $Ax = 0$ as a vector y satisfying this equation. They are actually
linearly independent vectors; it will be obvious once we introduce the set of all solutions in the next section. In fact, any solution of the homogeneous system is a linear
combination of these $\operatorname{null}(A)$ number of solutions.

We may further summarize our results for linear systems as follows.

Let a linear homogeneous system $Ax = 0$ have m equations and n unknowns.

1. If $\operatorname{rank}(A) = n$, then $Ax = 0$ has a unique solution, the trivial solution.
2. If $\operatorname{rank}(A) < n$, then $Ax = 0$ has infinitely many solutions.

Notice that if $m \geq n$, then both the cases are possible, whereas if $m < n$ then
$\operatorname{rank}(A) \leq m < n$; consequently, there must exist infinitely many solutions to the
homogeneous linear system.

For non-homogeneous linear systems, the same conclusion is drawn provided that
the system is consistent. To say it explicitly, let the linear system $Ax = b$ have m
equations and n unknowns.

1. If $\operatorname{rank}([A \mid b]) > \operatorname{rank}(A)$, then $Ax = b$ has no solutions.
2. If $\operatorname{rank}([A \mid b]) = \operatorname{rank}(A) = n$, then $Ax = b$ has a unique solution.
3. If $\operatorname{rank}([A \mid b]) = \operatorname{rank}(A) < n$, then $Ax = b$ has infinitely many solutions.

Notice that the number of equations plays no role in the nature of solutions, but
the rank of the system matrix, which is equal to the number of *linearly independent
equations*, is important.

Exercises for Sect. 2.4

1. Determine whether the following linear systems are consistent or not.

    (a) $2x_1 - x_2 - x_3 = 2, x_1 + x_3 - 4x_4 = 1, x_2 - x_3 - 4x_4 = 4$

(b) $2x_1 - x_2 - x_3 = 2, x_1 + x_3 - 4x_4 = 1, 3x_1 - x_2 - 4x_4 = 4$

2. For each of the following augmented matrices, determine whether the corresponding linear system has (i) no solutions, (ii) a unique solution, and (iii) infinitely many solutions.

(a) $\begin{bmatrix} 1 & -2 & 4 & | & 1 \\ 0 & 0 & 1 & | & 3 \\ 0 & 0 & 0 & | & 0 \end{bmatrix}$ (b) $\begin{bmatrix} 1 & -2 & 2 & | & -2 \\ 0 & 1 & -1 & | & 3 \\ 0 & 0 & 1 & | & 2 \end{bmatrix}$ (c) $\begin{bmatrix} 1 & -1 & 3 & | & 8 \\ 0 & 1 & 2 & | & 7 \\ 0 & 0 & 1 & | & 2 \end{bmatrix}$

3. In the following cases, what do you conclude about the number of solutions of the linear system $Ax = b$?

(a) $A \in \mathbb{F}^{5 \times 3}$ and $b = A_{\star 1} + A_{\star 2} = A_{\star 2} - A_{\star 3}$
(b) $A \in \mathbb{F}^{3 \times 4}$ and $b = A_{\star 1} + A_{\star 2} + A_{\star 3} + A_{\star 4}$
(c) $A \in \mathbb{F}^{3 \times 3}$ and $b = 3A_{\star 1} + 2A_{\star 2} + A_{\star 3} = 0$

4. Let A be a $7 \times n$ matrix of rank r, and let $b \in \mathbb{F}^{7 \times 1}$. For each of the following cases, determine, if possible, the number of solutions of the linear system $Ax = b$.
(a) $n = 8, r = 5$ (b) $n = 8, r = 7$ (c) $n = 6, r = 6$ (d) $n = 6, r = 5$

2.5 Gauss–Jordan Elimination

Gauss–Jordan elimination is an application of converting the augmented matrix to its row reduced echelon form for solving a linear system. We start with determining whether a system of linear equations has a solution or not. The consistency condition implies that if an entry in the b portion of the RREF of $[A \mid b]$ has become a pivot, then the system is inconsistent; otherwise, the system is consistent.

Example 2.8 Does the following system of linear equations have a solution?

$$5x_1 + 2x_2 - 3x_3 + x_4 = 7$$
$$x_1 - 3x_2 + 2x_3 - 2x_4 = 11$$
$$3x_1 + 8x_2 - 7x_3 + 5x_4 = 8$$

We take the augmented matrix and reduce it to its row reduced echelon form by elementary row operations.

$$\begin{bmatrix} 5 & 2 & -3 & 1 & | & 7 \\ 1 & -3 & 2 & -2 & | & 11 \\ 3 & 8 & -7 & 5 & | & 8 \end{bmatrix} \xrightarrow{R1} \begin{bmatrix} \boxed{1} & 2/5 & -3/5 & 1/5 & | & 7/5 \\ 0 & -17/5 & 13/5 & -11/5 & | & 48/5 \\ 0 & 34/5 & -26/5 & 22/5 & | & -19/5 \end{bmatrix}$$

$$\xrightarrow{R2} \begin{bmatrix} \boxed{1} & 0 & -5/17 & -1/17 & | & 43/17 \\ 0 & \boxed{1} & -13/17 & 11/17 & | & -48/17 \\ 0 & 0 & 0 & 0 & | & \boxed{77/5} \end{bmatrix}.$$

Here, $R_1 = E_{1/5}[1]$, $E_{-1}[2, 1]$, $E_{-3}[3, 1]$ and $R2 = E_{-5/17}[2]$, $E_{-2/5}[1, 2]$, $E_{-34/5}[3, 2]$.
Since an entry in the b portion has become a pivot, the system is inconsistent. In
fact, you can verify that the third row in A is simply first row minus twice the second
row, whereas the third entry in b is not the first entry minus twice the second entry.
Therefore, the system is inconsistent.   □

We write the **set of all solutions** of the system $Ax = b$ as $\mathrm{Sol}(A, b)$. That is,

$$\mathrm{Sol}(A, b) = \left\{ y \in \mathbb{F}^{n \times 1} : Ay = b \right\}.$$

As in Example 2.8, if there is no solution, then $\mathrm{Sol}(A, b) = \varnothing$.

Example 2.9 To illustrate the proof of Theorem 2.9, we change the last equation in
the previous example to make it consistent. We consider the new system

$$5x_1 + 2x_2 - 3x_3 + x_4 = 7$$
$$x_1 - 3x_2 + 2x_3 - 2x_4 = 11$$
$$3x_1 + 8x_2 - 7x_3 + 5x_4 = -15$$

Computation of the row reduced echelon form of the augmented matrix goes as
follows:

$$\begin{bmatrix} 5 & 2 & -3 & 1 & 7 \\ 1 & -3 & 2 & -2 & 11 \\ 3 & 8 & -7 & 5 & -15 \end{bmatrix} \xrightarrow{R1} \begin{bmatrix} \boxed{1} & 2/5 & -3/5 & 1/5 & 7/5 \\ 0 & -17/5 & 13/5 & -11/5 & 48/5 \\ 0 & 34/5 & -26/5 & 22/5 & -96/5 \end{bmatrix}$$

$$\xrightarrow{R2} \begin{bmatrix} \boxed{1} & 0 & -5/17 & -1/17 & 43/17 \\ 0 & \boxed{1} & -13/17 & 11/17 & -48/17 \\ 0 & 0 & 0 & 0 & 0 \end{bmatrix}.$$

Here, $R_1 = E_{1/5}[1]$, $E_{-1}[2, 1]$, $E_{-3}[3, 1]$ and $R2 = E_{-5/17}[2]$, $E_{-2/5}[1, 2]$, $E_{-34/5}[3, 2]$, as
earlier. The third row in the RREF is a zero row. Thus, the third equation is redundant.
Now, solving the new system in row reduced echelon form is easier. Writing as linear
equations, we have

$$\boxed{1}\,x_1 \qquad - \tfrac{5}{17}x_3 - \tfrac{1}{17}x_4 = \tfrac{43}{17}$$
$$\boxed{1}\,x_2 - \tfrac{13}{17}x_3 + \tfrac{11}{17}x_4 = -\tfrac{48}{17}$$

The unknowns corresponding to the pivots, that is, x_1 and x_2, are the basic variables,
and the other unknowns, x_3, x_4, are the free variables. The number of basic variables is
equal to the number of pivots, which is the rank of the system matrix. By assigning the
free variables x_i to any arbitrary values, say, α_i, the basic variables can be evaluated
in terms of α_i.
    We assign x_3 to α_3 and x_4 to α_4. Then, we have

$$x_1 = \frac{43}{17} + \frac{5}{17}\alpha_3 + \frac{1}{17}\alpha_4, \quad x_2 = -\frac{48}{17} + \frac{13}{17}\alpha_3 - \frac{11}{17}\alpha_4.$$

Therefore, any vector $y \in \mathbb{F}^{4\times1}$ in the form

$$y = \begin{bmatrix} \frac{43}{17} + \frac{5}{17}\alpha_3 + \frac{1}{17}\alpha_4 \\ -\frac{48}{17} + \frac{13}{17}\alpha_3 - \frac{11}{17}\alpha_4 \\ \alpha_3 \\ \alpha_4 \end{bmatrix} \quad \text{for } \alpha_3, \alpha_4 \in \mathbb{F}$$

is a solution of the linear system. Moreover, any solution of the linear system is in the above form. That is, the set of all solutions is given by

$$\text{Sol}(A, b) = \left\{ \begin{bmatrix} 43/17 \\ -48/17 \\ 0 \\ 0 \end{bmatrix} + \alpha_3 \begin{bmatrix} 5/17 \\ 13/17 \\ 1 \\ 0 \end{bmatrix} + \alpha_4 \begin{bmatrix} 1/17 \\ -11/17 \\ 0 \\ 1 \end{bmatrix} : \alpha_3, \alpha_4 \in \mathbb{F} \right\}.$$

Here, the vector $\begin{bmatrix} 43/17, -48/17, 0, 0 \end{bmatrix}^t$ is a particular solution of the original system. The two vectors $\begin{bmatrix} 5/17, 13/17, 1, 0 \end{bmatrix}^t$ and $\begin{bmatrix} 1/17, -11/17, 0, 1 \end{bmatrix}^t$ are linearly independent solutions of the corresponding homogeneous system. Notice that the nullity of the system matrix is 2. □

Instead of writing the RREF as a linear system again, we can reach at the set of all solutions quite mechanically. See the following procedure.

Gauss–Jordan Elimination

1. Reduce the augmented matrix $[A \mid b]$ to its RREF, say, $[A' \mid b']$.
2. If a pivot has appeared in b', then $Ax = b$ has no solutions.
3. Else, delete all zero rows from $[A' \mid b']$.
4. Insert zero rows in $[A' \mid b']$, if required, so that for each pivot, its row index is equal to its column index.
5. Insert zero rows at the bottom, if required, to make A' a square matrix. Call the updated matrix $[\tilde{A} \mid \tilde{b}]$.
6. Change the diagonal entries of the zero rows in $\tilde{A}$ from 0 to -1.
7. If the non-pivotal columns in $\tilde{A}$ are $u_1, \ldots, u_k$, then the set of all solutions is given by $\text{Sol}(A, b) = \{b' + \alpha_1 u_1 + \cdots + \alpha_k u_k\}$.

Example 2.10 We apply Gauss–Jordan elimination on the linear system of Example 2.9. The RREF of the augmented matrix as computed there is

$$[A' \mid b'] = \begin{bmatrix} \boxed{1} & 0 & -5/17 & -1/17 & 43/17 \\ 0 & \boxed{1} & -13/17 & 11/17 & -48/17 \\ 0 & 0 & 0 & 0 & 0 \end{bmatrix}.$$

We delete the zero row at the bottom. For each pivot, the row index is equal to its column index; so, no new zero row is to be inserted. Next, to make A' a square matrix, we adjoin two zero rows at the bottom. Next, we change the diagonal entries of all zero rows to -1. It yields the following matrix:

$$[\tilde{A} \mid \tilde{b}] = \begin{bmatrix} \boxed{1} & 0 & -5/17 & -1/17 & 43/17 \\ 0 & \boxed{1} & -13/17 & 11/17 & -48/17 \\ 0 & 0 & -1 & 0 & 0 \\ 0 & 0 & 0 & -1 & 0 \end{bmatrix}.$$

The non-pivotal columns are the third and the fourth columns. According to Gauss–Jordan elimination, the set of solutions is given by

$$\text{Sol}(A, b) = \left\{ \begin{bmatrix} 43/17 \\ -48/17 \\ 0 \\ 0 \end{bmatrix} + \alpha_1 \begin{bmatrix} -5/17 \\ -13/17 \\ -1 \\ 0 \end{bmatrix} + \alpha_2 \begin{bmatrix} -1/17 \\ 11/17 \\ 0 \\ -1 \end{bmatrix} : \alpha_1, \alpha_2 \in \mathbb{F} \right\}.$$

You may match this solution set with that in Example 2.9.   □

There are variations of Gauss–Jordan elimination. Instead of reducing the augmented matrix to its row reduced echelon form, if we reduce it to another intermediary form, called *the row echelon form*, then we obtain the method of **Gaussian elimination**. In the row echelon form, we do not require the entries above a pivot to be 0; also, the pivots need not be equal to 1. In that case, we will require back-substitution in solving a linear system. To illustrate this process, we redo Example 2.9 starting with the augmented matrix; it is as follows:

$$\begin{bmatrix} \boxed{5} & 2 & -3 & 1 & 7 \\ 1 & -3 & 2 & -2 & 11 \\ 3 & 8 & -7 & 5 & -15 \end{bmatrix} \xrightarrow{R_1} \begin{bmatrix} \boxed{5} & 2 & -3 & 1 & 7 \\ 0 & \boxed{-17/5} & 13/5 & -11/5 & 48/5 \\ 0 & 34/5 & -26/5 & 22/5 & -96/5 \end{bmatrix}$$

$$\xrightarrow{E_2[3,2]} \begin{bmatrix} \boxed{5} & 2 & -3 & 1 & 7 \\ 0 & \boxed{-17/5} & 13/5 & -11/5 & 48/5 \\ 0 & 0 & 0 & 0 & 0 \end{bmatrix}.$$

Here, $R_1 = E_{-1/5}[2, 1], E_{-3/5}[3, 1]$. The augmented matrix is now in row echelon form. It is a consistent system, since no entry in the b portion is a pivot. The pivots say that x_1, x_2 are basic variables and x_3, x_4 are free variables. We assign x_3 to α_3 and x_4 to α_4. Writing in equation form, we have

$$x_1 = 7 - 2x_2 + 3\alpha_3 - \alpha_4, \quad x_2 = -\tfrac{5}{17}\left(\tfrac{48}{5} - \tfrac{13}{5}\alpha_3 + \tfrac{11}{5}\alpha_4\right).$$

First we determine x_2 and then back-substitute. We obtain

$$x_1 = \tfrac{43}{17} + \tfrac{5}{17}\alpha_3 + \tfrac{1}{17}, \quad x_2 = -\tfrac{48}{17} + \tfrac{13}{17}\alpha_3 - \tfrac{11}{17}\alpha_4, \quad x_3 = \alpha_3, \quad x_4 = \alpha_4.$$

Thus, the solution set is given by

$$
\text{Sol}(A, b) = \left\{ \begin{bmatrix} 43/17 \\ -48/17 \\ 0 \\ 0 \end{bmatrix} + \alpha_3 \begin{bmatrix} 5/17 \\ 13/17 \\ 1 \\ 0 \end{bmatrix} + \alpha_4 \begin{bmatrix} 1/17 \\ -11/17 \\ 0 \\ 1 \end{bmatrix} : \alpha_3, \alpha_4 \in \mathbb{F} \right\}.
$$

As you see we end up with the same set of solutions as in Gauss–Jordan elimination.

Exercises for Sect. 2.5

1. Using Gauss–Jordan climination, and also Gaussian elimination, solve the following linear systems:

   (a) $3w + 2x + 2y - z = 2$, $2x + 3y + 4z = -2$, $y - 6z = 6$
   (b) $w + 4x + y + 3z = 1$, $2x + y + 3z = 0$, $w + 3x + y + 2z = 1$,
       $2x + y + 6z = 0$
   (c) $w - x + y - z = 1$, $w + x - y - z = 1$, $w - x - y + z = 2$,
       $4w - 2x - 2y = 1$

2. For each of the following augmented matrices, determine the solution set of the corresponding linear system:

   (a) $\begin{bmatrix} 1 & 0 & 0 & | & 2 \\ 0 & 0 & 1 & | & 1 \\ 0 & 0 & 1 & | & 3 \end{bmatrix}$ (b) $\begin{bmatrix} 1 & 4 & 0 & | & 2 \\ 0 & 0 & 1 & | & 3 \\ 0 & 0 & 0 & | & 1 \end{bmatrix}$ (c) $\begin{bmatrix} 1 & -3 & 0 & | & 2 \\ 0 & 0 & 1 & | & -2 \\ 0 & 0 & 0 & | & 0 \end{bmatrix}$

3. Let $B = \begin{bmatrix} 1 & 0 & 2 & 0 & -1 \\ 0 & 1 & 3 & 0 & -1 \\ 0 & 0 & 0 & 1 & 5 \\ 0 & 0 & 0 & 0 & 0 \end{bmatrix}$, $w = \begin{bmatrix} 3 \\ 2 \\ 0 \\ 2 \\ 0 \end{bmatrix}$, and let $b = \begin{bmatrix} 0 \\ 5 \\ 3 \\ 4 \end{bmatrix}$.

   Suppose A is a matrix such that $A_{*1} = [2, 1, -3, 2]^t$, $A_{*2} = [-1, 2, 3, 1]^t$, B is the RREF of A, and $Aw = b$. Determine the following:
   (a) $\text{Sol}(A, 0)$ (b) $\text{Sol}(A, b)$ (c) A

4. Let A be a matrix with $\text{Sol}(A, 0) = \{\alpha[1\ 0\ -1]^t + \beta[0\ 1\ 3]^t : \alpha, \beta \in \mathbb{R}\}$, and let $b = A_{*1} + 2A_{*2} + A_{*3}$. Determine $\text{Sol}(A, b)$.

5. Show that the linear system $x + y + kz = 1$, $x - y - z = 2$, $2x + y - 2z = 3$ has no solution for $k = 1$ and has a unique solution for each $k \neq 1$.

6. Determine, if possible, the values of a, b so that the following systems have (i) no solutions, (ii) unique solution, and (iii) infinitely many solutions:

   (a) $x_1 + 2x_2 + x_3 = 1$, $-x_1 + 4x_2 + 3x_3 = 2$, $2x_1 - 2x_2 + ax_3 = 3$
   (b) $x_1 + x_2 + 3x_3 = 2$, $x_1 + 2x_2 + 4x_3 = 3$, $x_1 + 3x_2 + ax_3 = b$

2.6 Problems

1. Let A be an $m \times n$ matrix of rank $r = \min\{m, n\}$. Let u, v be vectors such that
   $Au = Av$. In the following cases, can you conclude $u = v$? Explain.
   (a) $r = m$ (b) $r = n$ (c) $r < m$ (d) $r < n$
2. Let A_j be the matrix obtained from the identity matrix of order n by replacing the
   jth column by the vector $[b_1, \ldots, b_n]^t$. Using Cramer's rule, show that $\det(A_j) =$
   b_j for $1 \leq j \leq n$.
3. Let $A \in \mathbb{F}^{m \times n}$, and let $b \in \mathbb{F}^{m \times 1}$. Prove the following:

   (a) If the rows of A are linearly independent, then $Ax = b$ has at least one solution.
   (b) If the columns of A are linearly independent, then $Ax = b$ has at most one
   solution.

4. Let $A \in \mathbb{F}^{n \times n}$. Prove that the following are equivalent:

   (a) A is invertible.
   (b) $Ax = 0$ has no non-trivial solution.
   (c) $Ax = b$ has a unique solution for some $b \in \mathbb{F}^{n \times 1}$.
   (d) $Ax = b$ has at least one solution for each $b \in \mathbb{F}^{n \times 1}$.
   (e) $Ax = e_i$ has at least one solution for each $i \in \{1, \ldots, n\}$.
   (f) $Ax = b$ has at most one solution for each $b \in \mathbb{F}^{n \times 1}$.
   (g) $Ax = b$ has a unique solution for each $b \in \mathbb{F}^{n \times 1}$.
   (h) $\operatorname{rank}(A) = n$.
   (i) The RREF of A is I.
   (j) The rows of A are linearly independent.
   (k) The columns of A are linearly independent.
   (l) $\det(A) \neq 0$.
   (m) For each $B \in \mathbb{C}^{n \times n}$, $AB = 0$ implies that $B = 0$.
   (n) For each $B \in \mathbb{C}^{n \times n}$, $BA = 0$ implies that $B = 0$.

5. Let $A \in \mathbb{F}^{n \times n}$. Prove that $\operatorname{adj}(A)$ is invertible iff A is invertible. Further, if A is
   invertible, then show that $(\operatorname{adj}(A))^{-1} = \operatorname{adj}(A^{-1}) = (\det(A))^{-1}A$.
6. Let A be an $n \times n$ invertible matrix with $n > 1$. Show that $\det(\operatorname{adj}(A)) =$
   $(\det(A))^{n-1}$.
7. Show that if $\det(A) = 1$, then $\operatorname{adj}(\operatorname{adj}(A)) = A$.
8. Show that a consistent linear system cannot have n number of solutions, where
   $n > 1$ is a natural number.
9. Let $A, B \in \mathbb{F}^{m \times n}$ be in RREF. Show that if $Ax = 0$ and $Bx = 0$ have the same
   set of solutions, then $A = B$.
10. Let $A \in \mathbb{F}^{n \times n}$. Using Gaussian elimination, show that there exists a matrix P,
    which is a product of elementary matrices of Type 1, a lower triangular matrix L,
    and an upper triangular matrix U such that $A = PLU$.

Chapter 3
Matrix as a Linear Map

3.1 Subspace and Span

Recall that $\mathbb{F}$ stands for either $\mathbb{R}$ or $\mathbb{C}$; and $\mathbb{F}^n$ is either $\mathbb{F}^{1 \times n}$ or $\mathbb{F}^{n \times 1}$. Also, recall that a typical row vector in $\mathbb{F}^{1 \times n}$ is written as $[a_1, \ldots, a_n]$ and a column vector in $\mathbb{F}^{n \times 1}$ is written as $[a_1, \ldots, a_n]^t$. Both the row and column vectors are written uniformly as $(a_1, \ldots, a_n)$; these constitute the vectors in $\mathbb{F}^n$. In $\mathbb{F}^n$, we have a special vector, called the zero vector, which we denote by 0; that is, $0 = (0, \ldots, 0)$. And if $x = (a_1, \ldots, a_n) \in \mathbb{F}^n$, then its additive inverse is $-x = (-a_1, \ldots, -a_n)$.

The operations of addition and scalar multiplication in $\mathbb{F}^n$ enjoy the following properties:
For all $u, v, w \in \mathbb{F}^n$, and for all $\alpha, \beta \in \mathbb{F}$,

1. $u + v = v + u$.
2. $(u + v) + w = u + (v + w)$.
3. $u + 0 = 0 + u = u$.
4. $u + (-u) = -u + u = 0$.
5. $\alpha(\beta u) = (\alpha\beta)u$.
6. $\alpha(u + v) = \alpha u + \alpha v$.
7. $(\alpha + \beta)u = \alpha u + \beta u$.
8. $1u = u$.
9. $(-1)u = -u$.
10. If $u + v = u + w$, then $v = w$.
11. If $\alpha u = 0$, then $\alpha = 0$ or $u = 0$.

It so happens that the last three properties follow from the earlier ones. Any nonempty set where the two operations of addition and scalar multiplication are defined, and which enjoy the first eight properties above, is called a *vector space* over $\mathbb{F}$. In this sense, $\mathbb{F}^n$, that is, both $\mathbb{F}^{1 \times n}$ and $\mathbb{F}^{n \times 1}$, are vector spaces over $\mathbb{F}$. In such a general setting if a nonempty subset of a vector space is closed under both

© The Author(s), under exclusive license to Springer Nature Switzerland AG 2021
A. Singh, *Introduction to Matrix Theory*,
https://doi.org/10.1007/978-3-030-80481-7_3

the operations, then it is called a *subspace*. We may not need these general notions. However, we define a subspace of our specific vector spaces.

Let V be a nonempty subset of $\mathbb{F}^n$. We say that V is a **subspace of** $\mathbb{F}^n$ iff for each scalar $\alpha \in \mathbb{F}$ and for each pair of vectors $u, v \in V$, we have both $u + v \in V$ and $\alpha u \in V$.

This is the meaning of the informal phrase: V is closed under addition and scalar multiplication. It is easy to see that a nonempty subset V of $\mathbb{F}^n$ is a subspace of $\mathbb{F}^n$ iff the following single condition is satisfied:

for each $\alpha \in \mathbb{F}$ and for all vectors u, v from V, $\alpha u + v \in V$.

Example 3.1 1. $\{0\}$ and $\mathbb{F}^n$ are subspaces of $\mathbb{F}^n$.
2. Let $V = \{(a, b, c) : 2a + 3b + 5c = 0, a, b, c \in \mathbb{F}\}$. Clearly, $(0, 0, 0) \in V$. So, $V \neq \varnothing$. If $(a_1, b_1, c_1), (a_2, b_2, c_2) \in V$, then

$$2a_1 + 3b_1 + 5c_1 = 0, \quad 2a_2 + 3b_2 + 5c_2 = 0.$$

Adding them, we have $2(a_1 + a_2) + 3(b_1 + b_2) + 5(c_1 + c_2) = 0$. Therefore, $(a_1, b_1, c_1) + (a_2, b_2, c_2) \in V$.
If $\alpha \in \mathbb{F}$, then $2(\alpha a_1) + 3(\alpha b_1) + 5(\alpha c_1) = 0$. So, $\alpha(a_1, b_1, c_1) \in V$.
Hence, V is a subspace of $\mathbb{F}^3$.
3. Let $V = \{(a, b, c) : 2a + 3b + 5c = 1, a, b, c \in \mathbb{F}\}$. Clearly, $(1/2, 0, 0) \in V$. So, $V \neq \varnothing$. Also, $(0, 1/3, 0) \in V$.
We see that $(1/2, 0, 0) + (0, 1/3, 0) = (1/2, 1/3, 0)$. And

$$2 \times 1/2 + 3 \times 1/3 + 5 \times 0 = 2 \neq 1.$$

That is, $(1/2, 0, 0) + (0, 1/3, 0) \notin V$. Therefore, V is not a subspace of $\mathbb{F}^3$.
Also, notice that $2(1/2, 0, 0) \notin V$.
4. Let $\beta_1, \ldots, \beta_n \in \mathbb{F}$. Let $V = \{[a_1, \ldots, a_n] : \beta_1 a_1 + \cdots + \beta_n a_n = 0\}$. It is easy to check that V is a subspace of $\mathbb{F}^{1 \times n}$.
5. Let $\alpha, \beta_1, \ldots, \beta_n \in \mathbb{F}$ with $\alpha \neq 0$. Then, the subset $V = \{[a_1, \ldots, a_n] : \beta_1 a_1 + \cdots + \beta_n a_n = \alpha\}$ is not a subspace of $\mathbb{F}^{1 \times n}$. Why?   □

Verify that in a subspace V, all the properties (1–8) above hold true. This is the reason we call such a nonempty subset as a subspace. Further, if U and V are subspaces of $\mathbb{F}^n$ and $U \subseteq V$, then we say that U is a subspace of V.

A singleton with a nonzero vector is not a subspace. For example, $\{(1, 1)\}$ is not a subspace of $\mathbb{F}^2$ since $2(1, 1)$ is not an element of this set.

What about the set $\{\alpha(1, 1) : \alpha \in \mathbb{F}\}$? Take any two vectors from this set, say, $\alpha(1, 1)$ and $\beta(1, 1)$. Let $\gamma \in \mathbb{F}$. Now,

$$\gamma(\alpha(1, 1)) + \beta(1, 1) = (\gamma\alpha + \beta)(1, 1)$$

is an element of the set. Therefore, the set is a subspace of $\mathbb{F}^2$. Notice that this set is the set of all linear combinations of the vector $(1, 1)$.

Recall that a linear combination of vectors $v_1, \ldots, v_m$ is any vector in the form

$$\alpha_1 v_1 + \cdots + \alpha_m v_m$$

for scalars $\alpha_1, \ldots, \alpha_m$. If S is any nonempty subset of $\mathbb{F}^n$, we define span(S) as the set of all linear combinations of finite number of vectors from S. That is,

$$\text{span}(S) = \{\alpha_1 v_1 + \cdots + \alpha_m v_m : \alpha_1, \ldots, \alpha_m \in \mathbb{F}, v_1, \ldots, v_m \in S \text{ for an } m \in \mathbb{N}\}.$$

Also, we define span$(\varnothing) = \{0\}$. We read span(S) as **span** of S.

When $S = \{v_1, \ldots, v_m\}$, we also write span(S) as span$\{v_1, \ldots, v_m\}$. Thus,

$$\text{span}\{v_1, \ldots, v_m\} = \{\alpha_1 v_1 + \cdots + \alpha_m v_m : \alpha_1, \ldots, \alpha_m \in \mathbb{F}\}.$$

For instance, $v_1 + \cdots + v_m$ and $v_1 + 5v_2$ are in span$\{v_1, \ldots, v_m\}$. In the first case, each α_i is equal to 1, whereas in the second case, $\alpha_1 = 1, \alpha_2 = 5$ and all other α_is are 0.

Also, for $S \neq \varnothing$, we may write span(S) in the following way:

$$
\begin{aligned}
\text{span}(S) &= \cup_{m=1}^{\infty} \{\alpha_1 v_1 + \cdots + \alpha_m v_m : \alpha_1, \ldots, \alpha_m \in \mathbb{F}, v_1, \ldots, v_m \in S\} \\
&= \cup_{m=1}^{\infty} \text{span}\{v_1, \ldots, v_m\} \text{ for } v_1, \ldots, v_m \in S \\
&= \cup_{m=1}^{\infty} \text{span}(A) \text{ for } A \subseteq S \text{ with } |A| - m.
\end{aligned}
$$

Here, $|A|$ means the number of elements of the set A. Further, when we speak of a set of vectors, it is implicitly assumed that the set is a subset of $\mathbb{F}^n$ for some n.

Example 3.2 In $\mathbb{R}^2$, $U = \text{span}\{(1, 1)\} = \{(a, a) : a \in \mathbb{R}\}$ is subspace of $\mathbb{R}^2$; it is the straight line that passes through the origin with slope 1.

Again, let $V = \text{span}\{(1, 1), (1, 2)\} = \{(\alpha(1, 1) + \beta(1, 2) : \alpha, \beta \in \mathbb{R}\}$, in $\mathbb{R}^2$. Any vector in V can be written as $u + v$, where u is a vector on the straight line $y = x$, and v is vector on the straight line $y = 2x$. Here, a vector on the straight line means vector directed from the origin to a point on the straight line. It seems any vector in the plane can be written as $u + v$ for some such u and v. To show that this is the case, we try writing any point $(a, b) \in \mathbb{R}^2$ as $u + v$. So, suppose $(a, b) = \alpha(1, 1) + \beta(1, 2)$. Then, $a = \alpha + \beta, b = \alpha + 2\beta$. Solving, we obtain $\beta = b - a, \alpha = 2a - b$. We may now verify that

$$(a, b) = (2a - b)(1, 1) + (b - a)(1, 2).$$

Therefore, $\mathbb{R}^2 \subseteq V$. On the other hand, $V \subseteq \mathbb{R}^2$, since each linear combination of $(1, 1)$ and $(1, 2)$ is in $\mathbb{R}^2$. Therefore, $V = \mathbb{R}^2$.

Notice that both U and V are subspaces of $\mathbb{R}^2$. Further, U is a *proper subspace* of $\mathbb{R}^2$, whereas $V = \mathbb{R}^2$. $\qquad\square$

Theorem 3.1 Let $S \subseteq \mathbb{F}^n$. Then, $S \subseteq \text{span}(S)$, and span$(S)$ is a subspace of $\mathbb{F}^n$.

Proof Any vector $u \in S$ is a linear combination of vectors from S, for $u = 1u$. Thus, $S \subseteq \text{span}(S)$.

Now, if $S = \varnothing$, then $\text{span}(S) = \{0\}$. And if $S \neq \varnothing$, then $S \subseteq \text{span}(S)$ implies that $\text{span}(S) \neq \varnothing$. In any case, $\text{span}(S)$ is nonempty.

If $u, v \in \text{span}(S)$, then both of them are linear combinations of vectors from S. So, $\alpha u + v$ is also a linear combination of vectors from S for any scalar α. Therefore, $\text{span}(S)$ is a subspace of $\mathbb{F}^n$. ∎

We see that $\text{span}(\text{span}(S)) = \text{span}(S)$. (How?) In general, span of any subspace is the subspace itself.

Let V be a subspace of $\mathbb{F}^n$, and let $S \subseteq V$. We say that S is a **spanning subset of** V, or that S **spans** V iff $V = \text{span}(S)$. In this case, each vector in V can be expressed as a linear combination of vectors from S; we informally say that the vectors in S span V.

Notice that $(2, 2) \in \text{span}\{(1, 1)\}$. So, $\text{span}\{(1, 1), (2, 2)\} = \text{span}\{(1, 1)\}$. As in Example 3.2, we have in $\mathbb{F}^2$,

$$\text{span}\{(1, 1), (2, 2)\} = \text{span}\{(1, 1)\} \neq \mathbb{F}^2, \quad \text{span}\{(1, 1), (1, 2)\} = \mathbb{F}^2.$$

Further, the vectors $(1, 1), (1, 2), (1, -1)$ also span $\mathbb{F}^2$. In fact, since the first two vectors span $\mathbb{F}^2$, any list of vectors from $\mathbb{F}^2$ containing these two will also span $\mathbb{F}^2$.

Similarly, the vectors $e_1, \ldots, e_n$ in $\mathbb{F}^{n \times 1}$ span $\mathbb{F}^{n \times 1}$, where e_i is the column vector in $\mathbb{F}^{n \times 1}$ whose ith component is 1 and all other components are 0.

In this terminology, vectors $v_1, \ldots, v_n$ are linearly dependent iff one of the vectors in this list is in the span of the rest. If no vector in the list is in the span of the rest, then the vectors are linearly independent.

Exercises for Sect. 3.1

1. Let V be a nonempty subset of $\mathbb{F}^n$. Prove the following:

   (a) V is a subspace of $\mathbb{F}^n$ iff for all $\alpha, \beta \in \mathbb{F}$ and for all vectors $u, v \in V$, $\alpha u + \beta v \in V$.
   (b) V is a subspace of $\mathbb{F}^n$ iff V with addition as in $\mathbb{F}^n$ and scalar multiplication as in $\mathbb{F}^n$ satisfies the first eight properties of a vector space, as mentioned in the text.

2. Let U be a subspace of V, and let V be a subspace of W, where W is a subspace of $\mathbb{F}^n$. Is U a subspace of W?
3. Let $u, v_1, v_2, \ldots, v_n$ be $n + 1$ distinct vectors in $\mathbb{F}^n$, $S_1 = \{v_1, v_2, \ldots, v_n\}$, and let $S_2 = \{u, v_1, v_2, \ldots, v_n\}$. Prove that $\text{span}(S_1) = \text{span}(S_2)$ iff $u \in \text{span}(S_1)$.
4. Let A and B be subsets of $\mathbb{F}^n$. Prove or disprove the following:

   (a) A is a subspace of $\mathbb{F}^n$ if and only if $\text{span}(A) = A$.
   (b) If $A \subseteq B$, then $\text{span}(A) \subseteq \text{span}(B)$.
   (c) $\text{span}(A \cup B) = \{u + v : u \in \text{span}(A),\ v \in \text{span}(B)\}$.
   (d) $\text{span}(A \cap B) \subseteq \text{span}(A) \cap \text{span}(B)$.
   (e) $\text{span}(A) \cap \text{span}(B) \subseteq \text{span}(A \cap B)$.

5. Let $S \subseteq U \subseteq \mathbb{F}^n$, where U is a subspace of $\mathbb{F}^n$. Suppose for any subspace V of $\mathbb{F}^n$, if $S \subseteq V$ then $U \subseteq V$. Prove that $U = \text{span}(S)$.
6. Let A and B be subsets of $\mathbb{F}^n$. Prove or disprove: $A \cup B$ is linearly independent iff $\text{span}(A) \cap \text{span}(B) = \{0\}$.

3.2 Basis and Dimension

Suppose we are given with a subset S of a subspace V of $\mathbb{F}^n$. The subset S may or may not span V. If it spans V, it is possible that it has a proper subset which also spans V. For instance,

$$S = \{[1, 2, -3], [1, 0, -1], [2, -4, 2], [0, -2, 2]\}$$

spans the subspace $V = \{[a, b, c] : a + b + c = 0\}$ of $\mathbb{F}^{1 \times 3}$. Also, the subset

$$\{[1, 2, -3], [1, 0, -1], [2, -4, 2]\}$$

of S spans the same subspace V. Notice that S is linearly dependent. Reason:

$$[0, -2, 2] = (-1)[1, 2, -3] + [1, 0, -1].$$

On the other hand, the linearly independent subset $\{[1, 2, -3]\}$ of S does not span V. For instance,

$$[1, 0, 1] \in V, \quad [1, 0, -1] \neq \alpha[1, 2, -3], \quad \text{for any } \alpha \in \mathbb{F}.$$

That is, a spanning subset may be superfluous and a linearly independent set may be deficient. A linearly independent set which also spans a subspace may be just adequate in spanning the subspace.

Let V be a subspace of $\mathbb{F}^n$. Let B be a set of vectors from V. We say that the set B is a **basis** of V iff B is linearly independent and B spans V. Also, we define $\varnothing$ as the basis for the zero subspace $\{0\}$.

In what follows, we consider ordered sets, and the ordering of vectors in a set is shown by the way they are written. For instance, in the ordered set $\{v_1, v_2, v_3\}$, the vector v_1 is the first vector, v_2 is the second, and v_3 is the third, whereas in $\{v_2, v_3, v_1\}$, the vector v_2 is the first, v_3 is the second, and v_1 is the third. We assume implicitly that *each basis is an ordered set.*

It follows from Theorem 2.5 that each basis for a subspace of $\mathbb{F}^n$ has at most n vectors.

Example 3.3 1. It is easy to check that $B = \{e_1, \ldots, e_n\}$ is a basis of $\mathbb{F}^{n \times 1}$. Similarly, $E = \{e_1^t, \ldots, e_n^t\}$ is a basis of $\mathbb{F}^{1 \times n}$. These are the *standard bases* of the spaces.

2. We show that $B = \{[1, 2, -3], [1, 0, -1]\}$ is a basis of

$$V = \{[a, b, c] : a + b + c = 0, a, b, c \in \mathbb{F}\}.$$

First, $B \subseteq V$. Second, any vector in V is of the form $[a, b, -a - b]$ for $a, b \in \mathbb{F}$. Now,

$$[a, b, -a - b] = \tfrac{b}{2}[1, 2, -3] + \left(a - \tfrac{b}{2}\right)[1, 0, -1]$$

shows that $\mathrm{span}(B) = V$. For linear independence, suppose

$$\alpha[1, 2, -3] + \beta[1, 0, -1] = [0, 0, 0].$$

Then, $\alpha + \beta = 0, 2\alpha = 0, -3\alpha - \beta = 0$. It implies that $\alpha = \beta = 0$.
3. Also, $E = \{[1, -1, 0], [0, 1, -1]\}$ is another basis for the subspace V of $\mathbb{F}^{1 \times 3}$ in (2). $\square$

Let B be a basis of a subspace V of $\mathbb{F}^n$. If $C \subseteq V$ is any proper superset of B, then any vector in $C \backslash B$ is a linear combination of vectors from B. So, C is linearly dependent. On the other hand, if D is any proper subset of B, then each vector in $B \backslash D$ fails to be a linear combination of vectors from D. Otherwise, B would be linearly dependent. We thus say that

> A basis is a maximal linearly independent set.
> A basis is a minimal spanning set.

Of course, you can prove that a maximal linearly independent set is a basis, and a minimal spanning set is a basis. This would guarantee that each subspace of $\mathbb{F}^n$ has a basis. We take a more direct approach.

Theorem 3.2 *Each subspace of $\mathbb{F}^n$ has a basis with at most n vectors.*

Proof Let V be a subspace of $\mathbb{F}^n$. If $V = \{0\}$, then $\varnothing$ is a basis of V. Otherwise, choose a nonzero vector v_1 from V. Take $B = \{v_1\}$. If $V = \mathrm{span}(B)$, then B is a basis of V. Else, there exists a nonzero vector, say, $v_2 \in V \backslash \mathrm{span}(B)$. Update B to $B \cup \{v_2\}$. Notice that B is linearly independent. Continue this process to obtain larger and larger linearly independent sets in V. By Theorem 2.5, a linearly independent set in V cannot have more than n vectors. Thus, the process terminates with a basis of V having at most n vectors.   ∎

The zero subspace $\{0\}$ has a single basis $\varnothing$. But other subspaces do not have a unique basis. For instance, the subspace V in Example 3.3 has at least two bases. However, something remains same in all these bases. In that example, both the bases have exactly two vectors.

Theorem 3.3 *Let V be a subspace of $\mathbb{F}^n$. All bases of V have the same number of vectors.*

Proof Without loss of generality, let V be a subspace of $\mathbb{F}^{1\times n}$. By Theorem 2.5, a basis of V can have at most n vectors. So, let $\{u_1, \ldots, u_k\}$ and $\{v_1, \ldots, v_m\}$ be bases for V, where $k \leq n$ and $m \leq n$. Construct the matrix $A \in \mathbb{F}^{(k+m)\times n}$ by taking its rows as $u_1, \ldots, u_k, v_1, \ldots, v_m$ in that order.

The first k rows of A are linearly independent, and other rows are linear combinations of these k rows. Also, the last m rows of A are linearly independent and other rows are linear combinations of these m rows. By Theorem 2.6, $k = \operatorname{rank}(A) = m$. ∎

In view of Theorem 3.3, there exists a unique non-negative number associated with each subspace of $\mathbb{F}^n$, which is the number of vectors in any basis of the subspace.

Let V be a subspace of $\mathbb{F}^n$. The number of vectors in some (or any) basis for V is called the **dimension** of V. We write this number as $\dim(V)$ and also as $\dim V$.

Since $\{e_1, \ldots, e_n\}$ is a basis for $\mathbb{F}^{n\times 1}$, $\dim(\mathbb{F}^{n\times 1}) = n$. Similarly, $\dim(\mathbb{F}^{1\times n}) = n$. Remember that when we consider $\mathbb{C}^{n\times 1}$ or $\mathbb{C}^{1\times n}$, the scalars in any linear combination are complex numbers, and for $\mathbb{R}^{n\times 1}$ or $\mathbb{R}^{1\times n}$, the scalars are real numbers. Notice that $\dim(\{0\}) = \dim(\operatorname{span}(\varnothing)) = 0$; the dimension of any subspace of $\mathbb{F}^n$ is at most n.

Example 3.4 The subspace $U := \{[a, b, c, d] : a - 2b + 3c = 0 = d + a, a, b, c, d \in \mathbb{F}\}$ may be written as

$$U = \{[a, b, c, d] : [2b - 3c, b, c, -2b + 3c] : b, c \subset \mathbb{F}\}$$
$$= \{b[2, 1, 0, -2] + c[-3, 0, 1, 3] : b, c \in \mathbb{F}\}.$$

The vectors $[2, 1, 0, -2]$ and $[-3, 0, 1, 3]$ are linearly independent. Therefore, U has a basis $\{[2, 1, 0, -2], [-3, 0, 1, 3]\}$. So, $\dim(U) = 2$. ⊔

Recall that $|B|$ stands for the number of elements in a set B. For any subspace V of $\mathbb{F}^n$, and any subset B of $\mathbb{F}^n$, the following statements should be obvious:

1. If $|B| < \dim(V)$, then $\operatorname{span}(B)$ is a proper subspace of V.
2. If $|B| > \dim(V)$, then B is linearly dependent.
3. If $|B| = \dim(V)$ and $\operatorname{span}(B) = V$, then B is a basis of V.
4. If $|B| = \dim(V)$ and B is linearly independent, then B is a basis of V.
5. If U is a subspace of V, then $\dim(U) \leq \dim(V) \leq n$.
6. If B is a superset of a spanning set of V, then B is linearly dependent.
7. If B is a proper subset of a linearly independent subset of V, then B is linearly independent, and $\operatorname{span}(B)$ is a proper subspace of V.
8. Each spanning set of V contains a basis of V.
9. Each linearly independent subset of V can be extended to a basis of V.

For (8)–(9), we may employ the same construction procedure as in the proof of Theorem 3.2. A statement equivalent to (9) is proved below.

Theorem 3.4 (Basis Extension Theorem) *Let V be a subspace of $\mathbb{F}^n$. Each basis of a subspace of V can be extended to a basis of V.*

Proof Let B be a basis for a subspace U of V. If $U = V$, then B is a basis of V. Else, let $v_1 \in V \backslash U$. Now, $B \cup \{v_1\}$ is linearly independent. If this set spans V, then it is a basis of V. Otherwise, let $v_2 \in V \backslash \mathrm{span}(B \cup \{v_1\})$. If $B \cup \{v_1, v_2\}$ is not a basis of V, then let $v_3 \in V \backslash \mathrm{span}(B \cup \{v_1, v_2\})$. Continue this process. The process terminates since $\dim(V) \leq n$. Upon termination, we obtain a basis for V.   ∎

We can use elementary row operations to extract a basis for a subspace which is given in the form of span of some finite number of vectors. We write the vectors as row vectors, form a matrix A, and convert it to its RREF. Then, the pivotal rows of the RREF form the required basis. Also, those rows of A which have become the pivotal rows (monitoring row exchanges) form a basis.

Example 3.5 Let $U = \mathrm{span}\{(1, 1, 1, 1), (2, 1, 0, 3), (-1, 0, 1, -2), (0, 3, 2, 1)\}$. Find a basis for the subspace U of $\mathbb{F}^4$.

We start with the matrix with these vectors as its rows and convert it to its RREF as follows:

$$
\begin{bmatrix} 1 & 1 & 1 & 1 \\ 2 & 1 & 0 & 3 \\ -1 & 0 & 1 & -2 \\ 0 & 3 & 2 & 1 \end{bmatrix}
\xrightarrow{R1}
\begin{bmatrix} \boxed{1} & 1 & 1 & 1 \\ 0 & -1 & -2 & 1 \\ 0 & 1 & 2 & -1 \\ 0 & 3 & 2 & 1 \end{bmatrix}
\xrightarrow{R2}
\begin{bmatrix} \boxed{1} & 0 & -1 & 2 \\ 0 & \boxed{1} & 2 & -1 \\ 0 & 0 & 0 & 0 \\ 0 & 0 & -4 & 4 \end{bmatrix}
$$

$$
\xrightarrow{R3}
\begin{bmatrix} \boxed{1} & 0 & 0 & 1 \\ 0 & \boxed{1} & 0 & 1 \\ 0 & 0 & \boxed{1} & -1 \\ 0 & 0 & 0 & 0 \end{bmatrix}.
$$

Here, $R1 = E_{-2}[2, 1], E_1[3, 1]; \quad R2 = E_{-1}[2], E_{-1}[1, 2], E_{-1}[3, 2], E_{-3}[4, 2];$ and $R3 = E[3, 4], E_{-1/4}[3], E_1[1, 3], E_{-2}[2, 3]$.

The pivotal rows show that $\{(1, 0, 0, 1), (0, 1, 0, 1), (0, 0, 1, -1)\}$ is a basis for the given subspace. Notice that only one row exchange has been done in this reduction process, which means that the third row in the RREF corresponds to the fourth vector and the fourth row corresponds to the third vector. Thus, the pivotal rows correspond to the first, second, and the fourth vector, originally. This says that a basis for the subspace is also given by $\{(1, 1, 1, 1), (2, 1, 0, 3), (0, 3, 2, 1)\}$.

The reduction process confirms that the third vector is a linear combination of the first, second, and the fourth.   □

The method illustrated in the above example can be used in another situation. Suppose B is a basis for U, which is a subspace of $V \subseteq \mathbb{F}^n$. Assume that a basis E for V is also known. Then, we can extend B to a basis for V. Towards this, we may form a matrix with the vectors in B as row vectors and then place the vectors from

E as rows below those from B. Then, reduction of this matrix to RREF produces a basis for V, which is an extension of B.

Linear independence has something to do with invertibility of a square matrix. Suppose the rows of a matrix $A \in \mathbb{F}^{n \times n}$ are linearly independent. Then, the RREF of A has n number of pivots. That is, $\mathrm{rank}(A) = n$. Consequently, A is invertible. On the other hand, if a row of A is a linear combination of other rows, then this row appears as a zero row in the RREF of A. That is, A is not invertible.

Considering A^t instead of A, we conclude that A^t is invertible iff the rows of A are linearly independent. However, A is invertible iff A^t is invertible. Therefore, A is invertible iff its columns are linearly independent.

Theorem 3.5 *A square matrix is invertible iff its rows are linearly independent iff its columns are linearly independent.*

From Theorem 3.5, it follows that an $n \times n$ matrix is invertible iff its rows form a basis for $\mathbb{F}^{1 \times n}$ iff its columns form a basis for $\mathbb{F}^{n \times 1}$.

Exercises for Sect. 3.2

1. Let V be a subspace of $\mathbb{F}^n$, and let $B \subseteq V$. Prove the following:

    (a) B is a basis of V iff B is linearly independent and each proper superset of B is linearly dependent.
    (b) B is a basis of V iff B spans V and no proper subset of B spans V.

2. Prove statements in (1)–(7) listed after Example 3.4.
3. Let $\{x, y, z\}$ be a basis for a vector space V. Is $\{x + y, y + z, z + x\}$ also a basis for V?
4. Find a basis for the subspace $\{(a, b, c) \in \mathbb{R}^3 : a + b - 5c = 0\}$ of $\mathbb{R}^3$.
5. Find bases and dimensions of the following subspaces of $\mathbb{R}^5$:

    (a) $\{(a, b, c, d, e) \in \mathbb{R}^5 : a - c - d = 0\}$.
    (b) $\{(a, b, c, d, e) \in \mathbb{R}^5 : b = c = d, a + e = 0\}$.
    (c) $\mathrm{span}\{(1, -1, 0, 2, 1), (2, 1, -2, 0, 0), (0, -3, 2, 4, 2),$
    $(3, 3, -4, -2, -1), (5, 7, -3, -2, 0)\}$.

6. Extend the set $\{(1, 0, 1, 0), (1, 0, -1, 0)\}$ to a basis of $\mathbb{R}^4$.
7. Let $u_1, v, w, x_1, x_2, x_3, x_4 \in \mathbb{R}^5$ satisfy $x_1 = u + 2v + 3w$, $x_2 = 2u - 3v + 4w$, $x_3 = -3u + 4v - 5w$, and $x_4 = -u + 6v + w$. Is $\{x_1, x_2, x_3, x_4\}$ linearly dependent?
8. Prove that the only proper subspaces of $\mathbb{R}^2$ are the straight lines passing through the origin.
9. Describe all subspaces of $\mathbb{R}^3$.
10. Let $A \in \mathbb{F}^{n \times n}$. Let $\{v_1, \ldots, v_n\}$ be any basis of $\mathbb{F}^{n \times 1}$. Prove that A is invertible iff $Ax = v_i$ has at least one solution for each $i = 1, \ldots, n$.
11. Let A be a matrix. Call a row (column) of A *redundant* if it can be expressed as a linear combination of other rows (columns). Show the following:

(a) If a redundant row of A is deleted, then the column rank of A remains unchanged.
(b) If a redundant column of A is deleted, then the row rank of A remains unchanged.
(c) The row rank of A is equal to the column rank of A.

3.3 Linear Transformations

Let $A \in \mathbb{F}^{m \times n}$. If $x \in \mathbb{F}^{n \times 1}$, then $Ax \in \mathbb{F}^{m \times 1}$. Thus, we may view the matrix A as a function from $\mathbb{F}^{n \times 1}$ to $\mathbb{F}^{m \times 1}$ which maps x to Ax. That is, the map $A : \mathbb{F}^{n \times 1} \to \mathbb{F}^{m \times 1}$ is given by

$$A(x) = Ax \quad \text{for } x \in \mathbb{F}^{n \times 1}.$$

Notice that instead of using another symbol, we write the map obtained this way from a matrix A as A itself. Due to the properties of matrix product, the following are true for the map A:

1. $A(u + v) = A(u) + A(v)$ for all $u, v \in \mathbb{F}^{n \times 1}$.
2. $A(\alpha v) = \alpha A(v)$ for all $v \in \mathbb{F}^{n \times 1}$ and for all $\alpha \in \mathbb{F}$.

In this manner, a matrix is considered as a *linear transformation*. In fact, any function A from a vector space to another (both over the same field) satisfying the above two properties is called a linear transformation or a *linear map*.

To see the connection between the matrix as a rectangular array and as a function, consider the values of the matrix A at the standard basis vectors $e_1, \ldots, e_n$ in $\mathbb{F}^{n \times 1}$. The column vector $e_j \in \mathbb{F}^{n \times 1}$ has the jth entry as 1 and all other entries 0. Let $A = [a_{ij}] \in \mathbb{F}^{m \times n}$. The product Ae_j is a vector in $\mathbb{F}^{m \times 1}$, whose ith entry is

$$a_{i1} \cdot 0 + \cdots + a_{i(j-1)} \cdot 0 + a_{ij} \cdot 1 + a_{i(j+1)} \cdot 0 + \cdots + a_{in} \cdot 0 = a_{ij}.$$

That is, $Ae_j = [a_{1j}, \ldots, a_{ij}, \ldots, a_{mj}]^t = A_{\star j}$, the jth column of A.

Observation 3.1 A matrix $A \in \mathbb{F}^{m \times n}$ is viewed as the linear transformation $A : \mathbb{F}^{n \times 1} \to \mathbb{F}^{m \times 1}$, where $A(e_j) = A_{\star j}$, the jth column of A for $1 \leq j \leq n$, and $A(v) = Av$ for each $v \in \mathbb{F}^{n \times 1}$.

The *range* of the linear transformation A is the set $R(A) = \{Ax : x \in \mathbb{F}^{n \times 1}\}$. If $y \in R(A)$, then there exists an $x = [\alpha_1, \ldots, \alpha_n]^t \in \mathbb{F}^{n \times 1}$ such that $y = Ax$. The vector x can be written as $x = \alpha_1 e_1 + \cdots + \alpha_n e_n$. Thus, such a $y \in R(A)$ is written as

$$y = Ax = \alpha_1 Ae_1 + \cdots + \alpha_n Ae_n.$$

Conversely, each vector $\alpha_1 Ae_1 + \cdots + \alpha_n Ae_n = A(\alpha_1 e_1 + \cdots + \alpha_n e_n) \in R(A)$. Hence,

$$R(A) = \{\alpha_1 A e_1 + \cdots + \alpha_n A e_n : \alpha_1, \ldots, \alpha_n \in \mathbb{F}\}$$
$$= \{\alpha_1 A_{\star 1} + \cdots + \alpha_n A_{\star n} : \alpha_1, \ldots, \alpha_n \in \mathbb{F}\}$$
$$= \text{span of columns of } A.$$

Therefore, $R(A)$ is a subspace of $\mathbb{F}^{m \times 1}$. We refer to $R(A)$ as the **range space** of A. Since $R(A)$ is the span of the columns of A, it is also called the *column space of* A. Its dimension is the *column rank* of A. We know that

$$\dim(R(A)) = \text{ the column rank of } A = \text{rank}(A).$$

Similarly, the subspace of $\mathbb{F}^{1 \times n}$ which is spanned by the rows of A is called the *row space* of A. Notice that the nonzero rows in the RREF of A form a basis for the row space of A. The dimension of the row space is the *row rank* of A, which we know to be equal to $\text{rank}(A)$ also.

For an $m \times n$ matrix A, viewed as a linear transformation, the set of all vectors which map to the zero vector is denoted by $N(A)$. That is,

$$N(A) = \{x \in \mathbb{F}^{n \times 1} : Ax = 0\}.$$

We find that $N(A)$ is the set of all solutions of the linear system $Ax = 0$. Also,

if $u, v \in N(A)$ and $\alpha \in \mathbb{F}$, then $A(\alpha u + v) = \alpha A u + A v = 0$.

Therefore, $N(A)$ is a subspace of $\mathbb{F}^{n \times 1}$. We refer to $N(A)$ as the **null space** of A. The dimension of the null space is called the **nullity** of the matrix A and is denoted by $\text{null}(A)$. That is,

$$\text{null}(A) = \dim(N(A)).$$

Theorem 2.9 (4) implies that $\dim(R(A)) + \dim(N(A)) = n$. Since this will be used often, we mention it as a theorem. An alternate proof of this theorem is given in Problem 12.

Theorem 3.6 (Rank Nullity) *Let $A \in \mathbb{C}^{m \times n}$. Then, $\dim(R(A)) + \dim(N(A)) = \text{rank}(A) + \text{null}(A) = n$.*

Gauss–Jordan elimination process gives us the following mechanical way of construction of a basis for $N(A)$. First, we reduce A to its RREF B. Next, we throw away the zero rows at the bottom of B to obtain an $r \times n$ matrix C. If necessary, we insert zero rows in C to obtain an $n \times n$ matrix D so that the pivots become diagonal entries in D. Each diagonal entry of D which is on a non-pivotal column is now 0. We change each such 0 to -1. Call the new matrix as E. The non-pivotal columns in E form a basis for $N(A)$.

Example 3.6 Consider the system matrix in Example 2.9. We had its RREF with (boxed) pivots as shown below:

$$A = \begin{bmatrix} 5 & 2 & -3 & 1 \\ 1 & -3 & 2 & -2 \\ 3 & 8 & -7 & 5 \end{bmatrix} \longrightarrow \begin{bmatrix} \boxed{1} & 0 & -5/17 & -1/17 \\ 0 & \boxed{1} & -13/17 & 11/17 \\ 0 & 0 & 0 & 0 \end{bmatrix} = \text{RREF}(A).$$

The first two columns in RREF(A) are the pivotal columns. So, the first two columns in A form a basis for $R(A)$. That is,

a basis for $R(A)$ is $\{[5, 1, 3]^t, [2, -3, 8]^t\}$.

For a basis of $N(A)$, notice that each pivot has the row index equal to the column index; so, we do not require to insert zero rows between pivot rows. To make it a square matrix, we attach a zero row to the RREF at the bottom:

$$D = \begin{bmatrix} \boxed{1} & 0 & -5/17 & -1/17 \\ 0 & \boxed{1} & -13/17 & 11/17 \\ 0 & 0 & 0 & 0 \\ 0 & 0 & 0 & 0 \end{bmatrix}.$$

Then, we change the diagonal entries in the non-pivotal columns to -1. These changed non-pivotal columns form a basis for $N(A)$. That is,

a basis for $N(A)$ is $\left\{ \left[-\frac{5}{17}, -\frac{13}{17}, -1, 0 \right]^t, \left[-\frac{1}{17}, \frac{11}{17}, 0, -1 \right]^t \right\}$.

Thus, $\dim(R(A)) + \dim(N(A)) = 4 = $ the number of columns in A. $\square$

Example 3.7 Find bases for $R(A)$ and $N(A)$, where $A = \begin{bmatrix} 5 & 2 & -3 & 2 \\ 10 & 4 & -5 & 5 \\ -5 & -2 & 3 & -2 \end{bmatrix}$.

We reduce A to its RREF as in the following:

$$\begin{bmatrix} 5 & 2 & -3 & 2 \\ 10 & 4 & -5 & 5 \\ -5 & -2 & 3 & -2 \end{bmatrix} \longrightarrow \begin{bmatrix} \boxed{1} & 2/5 & -3/5 & 2/5 \\ 0 & 0 & 1 & 1 \\ 0 & 0 & 0 & 0 \end{bmatrix} \longrightarrow \begin{bmatrix} \boxed{1} & 2/5 & 0 & 1 \\ 0 & 0 & \boxed{1} & 1 \\ 0 & 0 & 0 & 0 \end{bmatrix}.$$

A basis for $R(A)$ is provided by the columns of A that correspond to the pivotal columns, that is, the first and the third columns of A. So,

a basis for $R(A) = \{[5, 10, -5]^t, [-3, -5, 3]^t\}$.

The second and fourth columns are linear combinations of these basis vectors, where the coefficients are provided by the entries in the non-pivotal columns of the RREF. That is, the second column of A is $\frac{2}{5}$ times the first column, and the fourth column is 1 times the first column plus 1 times the second column. Indeed, we may verify that

$$[2, 4, -2]^t = \tfrac{2}{5}[5, 10, -5]^t, \quad [2, 5, -2]^t = [5, 10, -5]^t + [-3, -5, 3]^t.$$

Towards computing a basis for $N(A)$, notice that the only zero row in the RREF is the third row. So, we delete it. Next, the pivot on the second row has the column index as 3. So, we insert a zero row between first and second rows. Next, we adjoin a zero row at the bottom to make it a square matrix. Finally, we change the diagonal entry of these new rows to -1. We then obtain the matrix

$$\begin{bmatrix} \boxed{1} & 2/5 & 0 & 1 \\ 0 & -1 & 0 & 0 \\ 0 & 0 & \boxed{1} & 1 \\ 0 & 0 & 0 & -1 \end{bmatrix}.$$

A basis for $N(A)$ is provided by the non-pivotal columns. That is,
a basis for $N(A) = \{[\tfrac{2}{5}, -1, 0, 0]^t, [1, 0, 1, -1]^t\}.$ □

Exercises for Sect. 3.3

1. Let $A = \begin{bmatrix} 1 & 2 & 1 & 1 & 1 \\ 3 & 5 & 3 & 4 & 3 \\ 1 & 1 & 1 & 2 & 1 \\ 5 & 8 & 5 & 7 & 5 \end{bmatrix}$. Write $r = \text{rank}(A)$ and $k = \text{null}(A)$.

   (a) Find bases for $R(A)$ and $N(A)$, and then determine r and k.
   (b) Express suitable $4 - r$ rows of A as linear combinations of other r rows.
   (c) Express suitable $5 - r$ columns of A as linear combinations of other r columns.

2. Construct bases for (i) the row space, (ii) the column space, and the null space for each of the following matrices:

   (a) $\begin{bmatrix} 1 & 2 & 4 \\ 3 & 1 & 7 \\ 2 & 4 & 8 \end{bmatrix}$ (b) $\begin{bmatrix} -3 & 1 & -3 \\ 1 & 2 & 8 \\ 3 & -1 & 4 \\ 4 & -2 & 2 \end{bmatrix}$ (c) $\begin{bmatrix} 3 & 4 & 5 & 6 \\ 2 & 1 & 3 & 2 \\ 1 & 3 & -2 & 1 \end{bmatrix}$

3. Let $A \in \mathbb{F}^{m \times n}$. Consider A as the linear transformation $A : \mathbb{F}^{n \times 1} \to \mathbb{F}^{m \times 1}$ given by $A(x) = Ax$ for each $x \in \mathbb{F}^{n \times 1}$. Prove the following:

   (a) A is one-one iff $N(A) = \{0\}$.
   (b) A is one-one and onto iff A maps any basis onto a basis of $\mathbb{F}^{n \times 1}$.
   (c) A is one-one iff A is onto.

4. Let $A \in \mathbb{F}^{m \times n}$. Let $P \in \mathbb{F}^{m \times m}$ be invertible. Is it true that the RREF of PA is same as the RREF of A?

5. Let A and B be matrices having the same RREF. Then which of the following may be concluded? Explain.
   (a) $R(A) = R(B)$ (b) $N(A) = N(B)$
   (c) The row space of A is equal to the row space of B.

6. For $A \in \mathbb{F}^{m \times n}$ and $B \in \mathbb{F}^{n \times k}$, prove: $\text{rank}(AB) \le \min\{\text{rank}(A), \text{rank}(B)\}$.

3.4   Coordinate Vectors

Let $B = \{v_1, \ldots, v_m\}$ be a basis of a subspace V of $\mathbb{F}^n$. Recall that a basis is assumed to be an ordered set of vectors. Let $v \in V$. As B spans V, the vector v is a linear combination of vectors from B. Can there be two distinct linear combinations? Suppose there exist scalars $a_1, \ldots, a_m, b_1, \ldots, b_m \in \mathbb{F}$ such that

$$v = a_1 v_1 + \cdots + a_m v_m = b_1 v_1 + \cdots + b_m v_m.$$

Then, $(a_1 - b_1)v_1 + \cdots + (a_m - b_m)v_m = 0$. Since B is linearly independent, $a_1 = b_1, \ldots, a_m = b_m$. That is, such a linear combination is unique.

Conversely, suppose that each vector in V is written uniquely as a linear combination of $v_1, \ldots, v_m$. To show linear independence of these vectors, suppose that

$$\alpha_1 v_1 + \cdots + \alpha_m v_m = 0$$

for scalars $\alpha_1, \ldots, \alpha_m$. We also have

$$0 v_1 + \cdots + 0 v_m = 0.$$

From the uniqueness of writing the zero vector as a linear combination of $v_1, \ldots, v_m$, we conclude that $\alpha_1 = 0, \ldots, \alpha_m = 0$. Therefore, B is linearly independent.

We note down the result we have proved.

Theorem 3.7 *Let $B = \{v_1, \ldots, v_m\}$ be an ordered set of vectors from a subspace V of $\mathbb{F}^n$. B is a basis of V iff for each $v \in V$, there exists a unique vector $[\alpha_1, \ldots, \alpha_m]^t$ in $\mathbb{F}^{m \times 1}$ such that $v = \alpha_1 v_1 + \cdots + \alpha_m v_m$.*

Notice that in Theorem 3.7, $\dim(V) = m$.

Let $B = \{v_1, \ldots, v_m\}$ be a basis for a subspace V of $\mathbb{F}^n$. Let $v \in V$. For $v = \alpha_1 v_1 + \cdots + \alpha_m v_m$, the unique column vector $[\alpha_1, \ldots, \alpha_m]^t \in \mathbb{F}^{m \times 1}$ is called the **coordinate vector** of v with respect to the basis B; it is denoted by $[v]_B$. In this sense, we say that a basis provides a coordinate system in a subspace. Notice that the ordering of basis vectors is important to obtain the coordinate vector.

Example 3.8 Let $V = \{[a, b, c]^t : a + 2b + 3c = 0, a, b, c \in \mathbb{R}\}$. This subspace has a basis
$$B = \{[0, 3, -2]^t, [2, -1, 0]^t\}.$$

For $v = [-3, 0, 1]^t \in V$, we have

$$v = [-3, 0, 1]^t = -\tfrac{1}{2}[0, 3, -2]^t - \tfrac{3}{2}[2, -1, 0]^t.$$

Therefore, $[v]_B = \left[-\tfrac{1}{2}, -\tfrac{3}{2} \right]^t$.

If $w \in V$ has coordinate vector given by $[w]_B = [1, 1]^t$, then

$$w = 1\,[0, 3, -2]^t + 1\,[2, -1, 0]^t = [2, 2, -2]^t.$$

Since $\dim(V) = 2$, all the coordinate vectors are elements of $\mathbb{F}^{2 \times 1}$. ☐

The coordinate vector induces a linear map from V onto $\mathbb{F}^{m \times 1}$. To see this, let $B = \{v_1, \ldots, v_m\}$ be a basis for a subspace V of $\mathbb{F}^n$. Suppose $u, v \in V$ have the coordinate vectors

$$[u]_B = [a_1, \ldots, a_m]^t, \quad [v]_B = [b_1, \ldots, b_m]^t.$$

It means $u = a_1 v_1 + \cdots a_m v_m$ and $v = b_1 v_1 + \cdots + b_m v_m$. Then,

$$\alpha u + v = (\alpha a_1 + b_1) + \cdots + (\alpha a_m + b_m) v_m.$$

That is, $[\alpha u + v]_B = \alpha [u]_B + [v]_B$. In a way, this is the *linear property* of the coordinate vector map. Also,

$$v_j = 0 \cdot v_1 + \cdots + 0 \cdot v_{j-1} + 1 \cdot v_j + 0 \cdot v_{j+1} + \cdots + 0 \cdot v_m.$$

Hence, $[v_j]_B = e_j$ for each $j \in \{1, \ldots, m\}$.

We summarize these simple facts about the coordinate vector.

Observation 3.2 Let $B = \{v_1, \ldots, v_m\}$ be a basis for a subspace V of $\mathbb{F}^n$. Let $\alpha_1, \ldots, \alpha_m \in \mathbb{F}$, and let $u_1, \ldots, u_m \in V$. Then,

1. $[\alpha_1 u_1 + \cdots + \alpha_m u_m]_B = \alpha_1 [u_1]_B + \cdots + \alpha_m [u_m]_B$.
2. $[v_j]_B = e_j$, the jth standard basis vector of $\mathbb{F}^{m \times 1}$ for $1 \leq j \leq m$.

Consider $\mathbb{F}^{n \times 1}$ as its own subspace. It has many bases. If E is the standard basis of $\mathbb{F}^{n \times 1}$ and $v = [a_1, \ldots, a_n]^t \in \mathbb{F}^{n \times 1}$, then $[v]_E = v$. When we change the basis E to another, say, B, then how does $[v]_B$ look like?

Theorem 3.8 Let $\{e_1, \ldots, e_n\}$ be the standard basis of $\mathbb{F}^{n \times 1}$, and let $B = \{u_1, \ldots, u_n\}$ be any basis of $\mathbb{F}^{n \times 1}$. Write $P = [u_1 \cdots u_n]$. Then, $P[e_j]_B = e_j = [u_j]_B$ for $1 \leq j \leq n$, and for each $u \in \mathbb{F}^{n \times 1}$, $P[u]_B = u$ and $[u]_B = P^{-1}u$.

Proof Let $j \in \{1, \ldots, n\}$. By Observation 3.2, $[u_j]_B = e_j$. Since u_j is the jth column of P, we have $u_j = Pe_j = P[u_j]_B$.

Let $u \in \mathbb{F}^{n \times 1}$. Then, $u = a_1 u_1 + \cdots + a_n u_n$ for unique scalars $a_1, \ldots, a_n$. Now,

$$\begin{aligned} P[u]_B &= P[a_1, \ldots, a_n]^t = P(a_1 e_1 + \cdots + a_n e_n) \\ &= a_1 Pe_1 + \cdots + a_n Pe_n = a_1 u_1 + \cdots + a_n u_n = u. \end{aligned}$$

As the columns of P form a basis for $\mathbb{F}^{n \times 1}$, the matrix P is invertible. Therefore, $[u]_B = P^{-1}u$. ∎

Example 3.9 Consider the basis $B = \{u_1, u_2, u_3\}$ for $\mathbb{R}^{3\times1}$, where $u_1 = [1, 1, 1]^t$, $u_2 = [1, 0, 1]^t$, and $u_3 = [1, 0, 0]^t$. The matrix P in Theorem 3.8 and its inverse are given by

$$P = [u_1 \ u_2 \ u_3] = \begin{bmatrix} 1 & 1 & 1 \\ 1 & 0 & 0 \\ 1 & 1 & 0 \end{bmatrix}, \quad P^{-1} = \begin{bmatrix} 0 & 1 & 0 \\ 0 & -1 & 1 \\ 1 & 0 & -1 \end{bmatrix}.$$

We see that

$$
\begin{aligned}
e_1 &= [1, 0, 0]^t = 0 \cdot u_1 + 0 \cdot u_2 + u_3 \\
e_2 &= [0, 1, 0]^t = u_1 - u_2 + 0 \cdot u_3 \\
e_3 &= [0, 0, 1]^t = 0 \cdot u_1 + u_2 - u_3
\end{aligned}
$$

Also, since $P[e_j]_B = e_j$, we have $[e_j]_B = P^{-1}e_j$. That is, the columns of P^{-1} are the coordinate vectors of e_j with respect to the basis B. Either way,

$$[e_1]_B = \begin{bmatrix} 0 \\ 0 \\ 1 \end{bmatrix}, \quad [e_2]_B = \begin{bmatrix} 1 \\ -1 \\ 0 \end{bmatrix}, \quad [e_3]_B = \begin{bmatrix} 0 \\ 1 \\ -1 \end{bmatrix}.$$

It leads to

$$
\begin{aligned}
P[e_1]_B &= P[0, 0, 1]^t = [1, 0, 0]^t = e_1 \\
P[e_2]_B &= P[1, -1, 0]^t = [0, 1, 0]^t = e_2 \\
P[e_3]_B &= P[0, 1, -1]^t = [0, 0, 1]^t = e_3
\end{aligned}
$$

That is, $P[e_j]_B = e_j$ for each j as stated in Theorem 3.8. We verify the other conclusion of the theorem for $u = [1, 2, 3]^t$. Here, $u = 2u_1 + u_2 - 2u_3$. Therefore,

$$[u]_B = \begin{bmatrix} 2 \\ 1 \\ -2 \end{bmatrix} = \begin{bmatrix} 0 & 1 & 0 \\ 0 & -1 & 1 \\ 1 & 0 & -1 \end{bmatrix}\begin{bmatrix} 1 \\ 2 \\ 3 \end{bmatrix} = P^{-1}u. \qquad \square$$

It raises a more general problem: if a vector v has the coordinate vector $[v]_B$ with respect to a basis B of $\mathbb{F}^{n\times1}$, and also a coordinate vector $[v]_C$ with respect to another basis C of $\mathbb{F}^{n\times1}$, then how are $[v]_B$ and $[v]_C$ related? We will address this question in Sect. 3.6.

Exercises for Sect. 3.4

In the following, a basis B for $\mathbb{F}^{3\times1}$ and a vector $u \in \mathbb{F}^{3\times1}$ are given. Compute the coordinate vector $[u]_B$.

1. $B = \{[0, 0, 1]^t, [1, 0, 0]^t, [0, 1, 0]^t\}, u = [1, 2, 3]^t$
2. $B = \{[0, 1, 1]^t, [1, 0, 1]^t, [1, 1, 0]^t\}, u = [2, 1, 3]^t$
3. $B = \{[1, 1, 1]^t, [1, 2, 1]^t, [1, 2, 3]^t\}, u = [3, 2, 1]^t$

3.5 Coordinate Matrices

Let $A \in \mathbb{F}^{m \times n}$. We view A as a linear transformation from $\mathbb{F}^{n \times 1}$ to $\mathbb{F}^{m \times 1}$. Let D and E be the standard bases of $\mathbb{F}^{n \times 1}$ and $\mathbb{F}^{m \times 1}$, respectively. Then, $[u]_D = u$ and $[Au]_E = Au$. We thus obtain

$$[Au]_E = A [u]_D.$$

If we choose a different pair of bases, that is, a basis B for $\mathbb{F}^{n \times 1}$ and a basis C for $\mathbb{F}^{m \times 1}$, then u has the coordinate vector $[u]_B$ and Au has the coordinate vector $[Au]_C$. Looking at the above equation, we ask:

Does there exist a matrix M such that $[Au]_C = M[u]_B$?
If so, is it unique?

The following theorem answers these questions.

Theorem 3.9 *Let $A \in \mathbb{F}^{m \times n}$. Let $B = \{u_1, \ldots, u_n\}$ and $C = \{v_1, \ldots, v_m\}$ be bases for $\mathbb{F}^{n \times 1}$ and $\mathbb{F}^{m \times 1}$, respectively. Then, there exists a unique matrix $M \in \mathbb{F}^{m \times n}$ such that $[Au]_C = M[u]_B$. Moreover, the entries of $M = [\alpha_{ij}]$ satisfy the following:*

$$Au_j = \alpha_{1j} v_1 + \cdots + \alpha_{mj} v_m \quad \text{for each } j \in \{1, \ldots, n\}.$$

Proof Let $j \in \{1, \ldots, n\}$. Since $Au_j \in \mathbb{F}^{m \times 1}$ and C is a basis for $\mathbb{F}^{m \times 1}$, there exist unique scalars $\alpha_{1j}, \ldots, \alpha_{mj}$ such that

$$Au_j = \alpha_{1j} v_1 + \cdots + \alpha_{mj} v_m.$$

Thus, we have unique mn scalars α_{ij} so that the above equation is true for each j. Construct the matrix $M = [\alpha_{ij}] \in \mathbb{F}^{m \times n}$. By Observation 3.2,

$$[Au_j]_C = \left[\alpha_{1j}, \ldots, \alpha_{mj}\right]^t = M e_j = M [u_j]_B \quad \text{for each } j \in \{1, \ldots, n\}.$$

We must verify that such an equality holds for each $u \in \mathbb{F}^{n \times 1}$. So, let $u \in \mathbb{F}^{n \times 1}$. As B is a basis of $\mathbb{F}^{n \times 1}$, there exist unique scalars $\beta_1, \ldots, \beta_n$ such that $u = \beta_1 u_1 + \cdots + \beta_n u_n$. Now,

$$
\begin{aligned}
[Au]_C &= [\beta_1 u_1 + \cdots + \beta_n u_n]_C = \beta_1 [Au_1]_C + \cdots + \beta_n [Au_n]_C \\
&= \beta_1 M [u_1]_B + \cdots + \beta_n M [u_n]_B = M [\beta_1 u_1 + \cdots + \beta_n u_n]_B = M [u]_B.
\end{aligned}
$$

This proves the existence of a required matrix M.

To prove uniqueness, let P and Q be matrices such that

$$[Au]_C = P [u]_B = Q [u]_B \quad \text{for each } u \in \mathbb{F}^{n \times 1}.$$

In particular, with $u = u_j$, we have

$$[Au_j]_C = P[u_j]_B = Q[u_j]_B \quad \text{for each } j \in \{1, \ldots, n\}.$$

Since $[u_j]_B = e_j$, we obtain $Pe_j = Qe_j$. That is, the jth column of P is equal to the jth column of Q for each $j \in \{1, \ldots, n\}$. Therefore, $P = Q$. ∎

In view of Theorem 3.9, we denote the matrix M as $[A]_{C,B}$ and give it a name. Let $A \in \mathbb{F}^{m \times n}$. Let B be a basis for $\mathbb{F}^{n \times 1}$, and let C be a basis for $\mathbb{F}^{m \times 1}$. The unique $m \times n$ matrix $[A]_{C,B}$ that satisfies

$$[Au]_C = [A]_{C,B}[u]_B \quad \text{for each } u \in \mathbb{F}^{n \times 1}$$

is called the **coordinate matrix** of A with respect to the bases B and C.

Theorem 3.9 provides a method to compute the coordinate matrix. Let $A \in \mathbb{F}^{m \times n}$, $B = \{u_1, \ldots, u_n\}$ be a basis for $\mathbb{F}^{n \times 1}$, and let $C = \{v_1, \ldots, v_n\}$ be a basis for $\mathbb{F}^{m \times 1}$. We express the image of each basis vector u_j in terms of the basis vector v_is as in the following:

$$
\begin{aligned}
Au_1 &= \alpha_{11}v_1 + \alpha_{21}v_2 + \cdots + \alpha_{m1}v_m \\
Au_2 &= \alpha_{12}v_1 + \alpha_{22}v_2 + \cdots + \alpha_{m2}v_m \\
&\;\;\vdots \\
Au_n &= \alpha_{1n}v_1 + \alpha_{2n}v_2 + \cdots + \alpha_{mn}v_m
\end{aligned}
\tag{3.1}
$$

Then, the coordinate matrix is (Mark which coefficient α_{ij} goes where)

$$
[A]_{C,B} =
\begin{bmatrix}
\alpha_{11} & \alpha_{12} & \cdots & \alpha_{1n} \\
\alpha_{21} & \alpha_{22} & \cdots & \alpha_{2n} \\
& & \vdots & \\
\alpha_{m1} & \alpha_{m2} & \cdots & \alpha_{mn}
\end{bmatrix}.
\tag{3.2}
$$

Also, notice that if the coordinate matrix is given by (3.2), then the equalities in (3.1) are satisfied.

Example 3.10 Compute the coordinate matrix $[A]_{C,B}$ of $A = \begin{bmatrix} 0 & 1 \\ 1 & 1 \\ 1 & -1 \end{bmatrix}$ with respect to the bases $B = \{\{[1, 2]^t, [3, 1]^t\}$ for $\mathbb{F}^{2 \times 1}$, and $C = \{[1, 0, 0]^t, [1, 1, 0]^t, [1, 1, 1]^t\}$ for $\mathbb{F}^{3 \times 1}$.

Write $u_1 = [1, 2]^t, u_2 = [3, 1]^t$ and $v_1 = [1, 0, 0]^t, v_2 = [1, 1, 0]^t, v_3 = [1, 1, 1]^t$. We obtain

$$
\begin{aligned}
Au_1 &= [2, 3, -1]^t &= -v_1 + 4v_2 - v_3 \\
Au_2 &= [1, 4, 2]^t &= -3v_1 + 2v_2 + 2v_3.
\end{aligned}
$$

Therefore, $[A]_{C,B} = \begin{bmatrix} -1 & -3 \\ 4 & 2 \\ -1 & 2 \end{bmatrix}$. $\qquad\qquad\qquad$ □

We use Theorem 3.8 in proving the following result, which shows another way for computing the coordinate matrix.

Theorem 3.10 *Let $A \in \mathbb{F}^{m \times n}$. Let $B = \{u_1, \ldots, u_n\}$ and $C = \{v_1, \ldots, v_m\}$ be bases for $\mathbb{F}^{n \times 1}$ and $\mathbb{F}^{m \times 1}$, respectively. Construct the matrices*

$$P = \begin{bmatrix} u_1 & \cdots & u_n \end{bmatrix}, \quad Q = \begin{bmatrix} v_1 & \cdots & v_m \end{bmatrix},$$

by taking the column vectors u_j and v_i as columns of the respective matrices. Then, $[A]_{C,B} = Q^{-1}AP$.

Proof Let $u = a_1 u_1 + \cdots + a_n u_n$. By Theorem 3.8, we obtain $P[u]_B = u$. Also, $Q[Au]_C = Au$.

Since the columns of $Q \in \mathbb{F}^{m \times m}$ are linearly independent, Q is invertible. Therefore, $[Au]_C = Q^{-1}Au = Q^{-1}AP[u]_B$.

It then follows that $[A]_{C,B} = Q^{-1}AP$. $\qquad\qquad\qquad$ ■

To summarize, if $A \in \mathbb{F}^{m \times n}$, B is a basis for $\mathbb{F}^{n \times 1}$, and C is a basis for $\mathbb{F}^{m \times 1}$, then the coordinate matrix $[A]_{C,B}$ can be obtained in two ways.

Due to Theorem 3.9, we express the A-image of the basis vectors in B as linear combinations of basis vectors in C. The coefficients of each such image form the corresponding columns of the matrix $[A]_{C,B}$.

Alternatively, using Theorem 3.10, we construct P as the matrix whose jth column is the jth vector in B and the matrix Q by taking its ith column as the ith vector in C. Then, we have $[A]_{C,B} = Q^{-1}AP$.

Since inverse of a matrix is computed using the reduction to RREF, the coordinate matrix may be computed the same way. If $B = \{u_1, \ldots, u_n\}$ and $C = \{v_1, \ldots, v_m\}$ are bases for $\mathbb{F}^{n \times 1}$ and $\mathbb{F}^{m \times 1}$, respectively, then

$$P = \begin{bmatrix} u_1 & \cdots & u_n \end{bmatrix}, \quad Q = \begin{bmatrix} v_1 & \cdots & v_m \end{bmatrix}.$$

Now, $AP = \begin{bmatrix} Au_1 & \cdots & Au_n \end{bmatrix}$. To compute $Q^{-1}AP$, we start with the augmented matrix $[Q \mid AP]$ and then proceed towards its RREF. Since Q^{-1} exists, the RREF will look like $[I \mid Q^{-1}AP]$. Schematically, the computation may be written as in the following:

$$\begin{bmatrix} v_1 & \cdots & v_m \mid Au_1 & \cdots & Au_n \end{bmatrix} \xrightarrow{RREF} [I \mid [A]_{C,B}].$$

Example 3.11 Compute the coordinate matrix $[A]_{C,B}$ of $A = \begin{bmatrix} 0 & 1 \\ 1 & 1 \\ 1 & -1 \end{bmatrix}$ with respect to the bases $B = \{[1, 2]^t, [3, 1]^t\}$ for $\mathbb{F}^{2 \times 1}$, and $C = \{[1, 0, 0]^t, [1, 1, 0]^t, [1, 1, 1]^t\}$ for $\mathbb{F}^{3 \times 1}$.

Writing the vectors of B as u_1 and u_2 in that order, we have

$$Au_1 = \begin{bmatrix} 0 & 1 \\ 1 & 1 \\ 1 & -1 \end{bmatrix} \begin{bmatrix} 1 \\ 2 \end{bmatrix} = \begin{bmatrix} 2 \\ 3 \\ -1 \end{bmatrix}, \quad Au_2 = \begin{bmatrix} 0 & 1 \\ 1 & 1 \\ 1 & -1 \end{bmatrix} \begin{bmatrix} 3 \\ 1 \end{bmatrix} = \begin{bmatrix} 1 \\ 4 \\ 2 \end{bmatrix}.$$

We then convert the augmented matrix of columns as vectors from C followed by Au_1 and Au_2 to RREF. It is as follows:

$$\begin{bmatrix} 1 & 1 & 1 & 2 & 1 \\ 0 & 1 & 1 & 3 & 4 \\ 0 & 0 & 1 & -1 & 2 \end{bmatrix} \xrightarrow{R1} \begin{bmatrix} 1 & 0 & 0 & -1 & -3 \\ 0 & 1 & 0 & 4 & 2 \\ 0 & 0 & 1 & -1 & 2 \end{bmatrix}.$$

Here, $R1 = E_{-1}[1, 2]$, $E_{-1}[2, 3]$. Therefore, $[A]_{C,B} = \begin{bmatrix} -1 & -3 \\ 4 & 2 \\ -1 & 2 \end{bmatrix}$.

It is easy to verify that $Au_1 = -v_1 + 4v_2 - v_3$ and $Au_2 = -3v_1 + 2v_2 + 2v_3$ as required by (3.1)–(3.2). □

The linear property of coordinate vectors can be extended to coordinate matrices; see Exercise 2. Moreover, the product formula $[Au]_C = [A]_{C,B}[u]_B$ has a similar looking expression for product of matrices.

Theorem 3.11 *Let $A \in \mathbb{F}^{m \times k}$ and $B \in \mathbb{F}^{k \times n}$ be matrices. Let C, D and E be bases for $\mathbb{F}^{n \times 1}$, $\mathbb{F}^{k \times 1}$ and $\mathbb{F}^{m \times 1}$, respectively. Then, $[AB]_{E,C} = [A]_{E,D}[B]_{D,C}$.*

Proof For each $v \in \mathbb{F}^{n \times 1}$,

$$[A]_{E,D}[B]_{D,C}[v]_C = [A]_{E,D}[Bv]_D = [ABv]_E = [AB]_{E,C}[v]_C.$$

Therefore, $[A]_{E,D}[B]_{D,C} = [AB]_{E,C}$. ∎

Notice that in Theorem 3.11, the matrices are the linear transformations

$$A : \mathbb{F}^{k \times 1} \to \mathbb{F}^{m \times 1}, \quad B : \mathbb{F}^{n \times 1} \to \mathbb{F}^{k \times 1}, \quad AB : \mathbb{F}^{n \times 1} \to \mathbb{F}^{m \times 1}.$$

Since $ABv = (AB)(v) = A(Bv)$, the linear transformation AB is the composition of the maps A and B. The composition map AB transforms a vector from $\mathbb{F}^{n \times 1}$ to one in $\mathbb{F}^{k \times 1}$ by using B first, and then it uses A to transform that vector from $\mathbb{F}^{k \times 1}$ to one in $\mathbb{F}^{m \times 1}$. Our result says that the coordinate matrix of the composition map AB is simply the product of the coordinate matrices of the maps A and B.

Exercises for Sect. 3.5

1. Let $A = \begin{bmatrix} 1 & 1 & 1 \\ 1 & 2 & 2 \\ 1 & 2 & 3 \end{bmatrix}$. Consider the bases $B = \{[1, 0, 1]^t, [0, 1, 1]^t, [1, 1, 0]^t$ and $C = \{[1, 1, 1]^t, [1, 2, 1]^t, [1, 1, 0]^t\}$ for $\mathbb{F}^{3 \times 1}$. Compute $[A]_{C,B}$ and $[A]_{B,C}$ in the following ways:

    (a) By expressing A-images of the basis vectors of B in terms of those of C and vice versa.

    (b) By computing the inverses of two matrices constructed from the basis vectors.

    (c) By using the RREF conversion as mentioned in the text.

2. Let $A, B \in \mathbb{F}^{m \times n}$, C be a basis of $\mathbb{F}^{n \times 1}$, and D be a basis for $\mathbb{F}^{m \times 1}$, and let $\alpha \in \mathbb{F}$. Show that $[\alpha A + B]_{D,C} = \alpha [A]_{D,C} + [B]_{D,C}$.

3. Take a 2×3 real matrix A and a 3×2 real matrix B. Also, take the bases $C = \{[1, 1], [0, 1]\}$, $D = \{[1, 0, 0], [0, 1, 0], [0, 0, 1]\}$ and $E = \{[1, 0], [1, 1]\}$ for the spaces $\mathbb{R}^{2 \times 1}$, $\mathbb{R}^{3 \times 1}$ and $\mathbb{R}^{2 \times 1}$, respectively. Verify Theorem 3.11.

4. Construct a matrix $A \in \mathbb{R}^{2 \times 2}$, a vector $v \in \mathbb{R}^{2 \times 1}$, and a basis $B = \{u_1, u_2\}$ for $\mathbb{R}^{2 \times 1}$ satisfying $[Av]_B \neq A[v]_B$.

3.6 Change of Basis Matrix

Recall that $[A]_{C,B}$ is that matrix in $\mathbb{F}^{m \times n}$ which satisfies the equation:

$$[Au]_C = [A]_{C,B}[u]_B \quad \text{for each } u \in \mathbb{F}^{n \times 1}.$$

Further, $[A]_{C,B}$ is given by $Q^{-1}AP$, where P is the matrix whose jth column is the jth basis vector of B, and Q is the matrix whose ith column is the ith basis vector of C. In particular, taking $A \in \mathbb{F}^{n \times n}$ as the identity matrix, writing B as O, and C as N, we see that

$$[u]_N = [I]_{N,O}[u]_O = Q^{-1}P[u]_O \quad \text{for each } u \in \mathbb{F}^{n \times 1}. \tag{3.3}$$

In (3.3), P is the matrix whose jth column is the jth basis vector of O, and Q is the matrix whose ith column is the ith basis vector of N for $1 \leq i, j \leq n$. Both O and N are bases for $\mathbb{F}^{n \times 1}$.

The formula in (3.3) shows how the coordinate vector changes when a basis changes. The matrix $I_{N,O}$, which is equal to $Q^{-1}P$, is called the **change of basis matrix** or the **transition matrix** when the basis changes from an old basis O to a new basis N.

Observe that the change of basis matrix $I_{N,O}$ can also be computed by expressing the basis vectors of O as linear combinations of basis vectors of N as stipulated by Theorem 3.9, or equivalently, in (3.1)–(3.2). Alternatively, if $O = \{u_1, \ldots, u_n\}$ and $N = \{v_1, \ldots, v_n\}$, the change of basis matrix $I_{N,O}$, which is equal to $Q^{-1}P$, may be computed by using reduction to RREF. Schematically, it is given as follows:

$$O = \{u_1, \ldots, u_n\}, N = \{v_1, \ldots, v_n\}, \quad \begin{bmatrix} v_1 & \cdots & v_n \mid u_1 & \cdots & u_n \end{bmatrix} \xrightarrow{RREF} [I \mid I_{N,O}].$$

Example 3.12 Consider the following bases for $\mathbb{R}^{3 \times 1}$:

$$O = \{[1, 0, 1]^t, [1, 1, 0]^t, [0, 1, 1]^t\}, \quad N = \{[1, -1, 1]^t, [1, 1, -1]^t, [-1, 1, 1]^t\}.$$

Find the change of basis matrix $I_{N,O}$ when the basis changes from O to N. Also, find the coordinate matrix $[A]_{N,O}$ of the matrix $A = \begin{bmatrix} 1 & 1 & 1 \\ -1 & 0 & 1 \\ 0 & 1 & 0 \end{bmatrix}$ with respect to the pair of bases O and N. Verify that

$$\begin{bmatrix} 1 \\ 2 \\ 3 \end{bmatrix}_N = I_{N,O} \begin{bmatrix} 1 \\ 2 \\ 3 \end{bmatrix}_O, \quad \left[A \begin{bmatrix} 1 \\ 2 \\ 3 \end{bmatrix} \right]_N = [A]_{N,O} \begin{bmatrix} 1 \\ 2 \\ 3 \end{bmatrix}_O.$$

To solve this problem, we express the basis vectors of O as linear combinations of those of N, as in Theorem 3.9. The details are as follows:

$$\begin{aligned}
[1, 0, 1]^t &= 1[1, -1, 1]^t + \tfrac{1}{2}[1, 1, -1]^t + \tfrac{1}{2}[-1, 1, 1]^t \\
[1, 1, 0]^t &= \tfrac{1}{2}[1, -1, 1]^t + 1[1, 1, -1]^t + \tfrac{1}{2}[-1, 1, 1]^t \\
[0, 1, 1]^t &= \tfrac{1}{2}[1, -1, 1]^t + \tfrac{1}{2}[1, 1, -1]^t + 1[-1, 1, 1]^t.
\end{aligned}$$

The coefficients from the first equation give the first column of $I_{N,O}$ and so on. Thus, we obtain

$$I_{N,O} = \begin{bmatrix} 1 & 1/2 & 1/2 \\ 1/2 & 1 & 1/2 \\ 1/2 & 1/2 & 1 \end{bmatrix}.$$

Alternatively, by Theorem 3.10, the change of basis matrix is given by

$$I_{N,O} = \begin{bmatrix} 1 & 1 & -1 \\ -1 & 1 & 1 \\ 1 & -1 & 1 \end{bmatrix}^{-1} \begin{bmatrix} 1 & 1 & 0 \\ 0 & 1 & 1 \\ 1 & 0 & 1 \end{bmatrix} = \begin{bmatrix} 1/2 & 0 & 1/2 \\ 1/2 & 1/2 & 0 \\ 0 & 1/2 & 1/2 \end{bmatrix} \begin{bmatrix} 1 & 1 & 0 \\ 0 & 1 & 1 \\ 1 & 0 & 1 \end{bmatrix} = \begin{bmatrix} 1 & 1/2 & 1/2 \\ 1/2 & 1 & 1/2 \\ 1/2 & 1/2 & 1 \end{bmatrix}.$$

Again, using the RREF conversion we may obtain $I_{N,O}$ as follows:

$$\begin{bmatrix} 1 & 1 & -1 & | & 1 & 1 & 0 \\ -1 & 1 & 1 & | & 0 & 1 & 1 \\ 1 & -1 & 1 & | & 1 & 0 & 1 \end{bmatrix} \xrightarrow{R1} \begin{bmatrix} 1 & 1 & -1 & | & 1 & 1 & 0 \\ 0 & 2 & 0 & | & 1 & 2 & 1 \\ 0 & -2 & 2 & | & 0 & -1 & 1 \end{bmatrix}$$

$$\xrightarrow{R2} \begin{bmatrix} 1 & 0 & -1 & | & 1/2 & 0 & -1/2 \\ 0 & 1 & 0 & | & 1/2 & 1 & 1/2 \\ 0 & 0 & 2 & | & 1 & 1 & 2 \end{bmatrix} \xrightarrow{R3} \begin{bmatrix} 1 & 0 & 0 & | & 1 & 1/2 & 1/2 \\ 0 & 1 & 0 & | & 1/2 & 1 & 1/2 \\ 0 & 0 & 1 & | & 1/2 & 1/2 & 1 \end{bmatrix}.$$

Here, $R1 = E_1[2, 1], E_{-1}[3, 1]$; $R2 = E_{1/2}[2], E_{-1}[1, 2], E_2[3, 2]$; $R3 = E_{1/2}[3], E_1[1, 3]$. The matrix in the RREF, to the right of the vertical bar, is the required change of basis matrix $I_{N,O}$.

To verify our result for $[1, 2, 3]^t$, notice that

$$[1, 2, 3]^t = 1[1, 0, 1]^t + 0[1, 1, 0]^t + 2[0, 1, 1]^t$$
$$[1, 2, 3]^t = 2[1, -1, 1]^t + \tfrac{3}{2}[1, 1, -1]^t + \tfrac{5}{2}[-1, 1, 1]^t.$$

Therefore, $[[1, 2, 3]^t]_O = [1, 0, 2]^t$ and $[[1, 2, 3]^t]_N = [2, 3/2, 5/2]^t$. Then,

$$I_{N,O} \begin{bmatrix} 1 \\ 2 \\ 3 \end{bmatrix}_O = \begin{bmatrix} 1 & 1/2 & 1/2 \\ 1/2 & 1 & 1/2 \\ 1/2 & 1/2 & 1 \end{bmatrix} \begin{bmatrix} 1 \\ 0 \\ 2 \end{bmatrix} = \begin{bmatrix} 2 \\ 3/2 \\ 5/2 \end{bmatrix} = \begin{bmatrix} 1 \\ 2 \\ 3 \end{bmatrix}_N .$$

According to Theorem 3.10,

$$[A]_{N,O} = \begin{bmatrix} 1 & 1 & -1 \\ -1 & 1 & 1 \\ 1 & -1 & 1 \end{bmatrix}^{-1} A \begin{bmatrix} 1 & 1 & 0 \\ 0 & 1 & 1 \\ 1 & 0 & 1 \end{bmatrix} = \tfrac{1}{2} \begin{bmatrix} 2 & 3 & 3 \\ 2 & 1 & 3 \\ 0 & 0 & 2 \end{bmatrix} .$$

As to the verification,

$$A \begin{bmatrix} 1 \\ 2 \\ 3 \end{bmatrix} = \begin{bmatrix} 6 \\ 2 \\ 6 \end{bmatrix} = 4 \begin{bmatrix} 1 \\ -1 \\ 1 \end{bmatrix} + 4 \begin{bmatrix} 1 \\ 1 \\ -1 \end{bmatrix} + 2 \begin{bmatrix} -1 \\ 1 \\ 1 \end{bmatrix} .$$

We find that $[A]_{N,O} \begin{bmatrix} 1 \\ 2 \\ 3 \end{bmatrix}_O = [A]_{N,O} \begin{bmatrix} 1 \\ 0 \\ 2 \end{bmatrix} = \begin{bmatrix} 4 \\ 4 \\ 2 \end{bmatrix} = \begin{bmatrix} A \begin{bmatrix} 1 \\ 2 \\ 3 \end{bmatrix} \end{bmatrix}_N .$ $\qquad\square$

Exercises for Sect. 3.4

Let $\quad A = \begin{bmatrix} 3 & 1 & 0 \\ 0 & 1 & 3 \\ -1 & -1 & -2 \end{bmatrix}, \quad V = \{[a, b, c]^t : a + 2b + 3c = 0, \ a, b, c \in \mathbb{R}\},$
$B = \{[0, 3, -2]^t, [2, -1, 0]^t\}$, and let $C = \{[1, 1, -1]^t, [3, 0, -1]^t\}$.

1. Show that V is a two-dimensional subspace of $\mathbb{R}^{3\times 1}$.
2. Show that $A : V \to V$ defined by $A(v) = Av$ for $v \in V$ is a well-defined function.
3. Show that B and C are bases for V.
4. Extend the basis B of V to a basis O for $\mathbb{F}^{3\times 1}$.
5. Extend the basis C of V to a basis N for $\mathbb{F}^{3\times 1}$.
6. Find the change of basis matrix $I_{N,O}$.
7. Verify that $[v]_N = I_{N,O} [v]_O$ for $v = [-4, -1, 2]^t$.
8. Find the matrix $[A]_{N,O}$.
9. Verify that $[Av]_N = [A]_{N,O} [v]_O$ for $v = [-4, -1, 2]^t$.

3.7 Equivalence and Similarity

In view of Theorem 3.10, two matrices $A, B \in \mathbb{F}^{m \times n}$ are called **equivalent** iff there exist invertible matrices $P \in \mathbb{F}^{n \times n}$ and $Q \in \mathbb{F}^{m \times m}$ such that $B = Q^{-1}AP$.

Notice that 'is equivalent to' is an equivalence relation on $\mathbb{F}^{m \times n}$. Equivalent matrices represent the same matrix (linear transformation) with respect to possibly different pairs of bases.

From Theorem 2.8, it follows that ranks of two equivalent matrices are equal.

We can construct a matrix of rank r relatively easily. Let $r \leq \min\{m, n\}$. The matrix $R_r \in \mathbb{F}^{m \times n}$ whose first r columns are the standard basis vectors $e_1, \ldots, e_r$ of $\mathbb{F}^{m \times 1}$ and all other columns are zero columns has rank r. This matrix, in block form, looks as follows:

$$R_r = \begin{bmatrix} I_r & 0 \\ 0 & 0 \end{bmatrix}.$$

Such a matrix is called a **rank echelon** matrix. If R_r is an $m \times n$ matrix, then necessarily $r \leq \min\{m, n\}$. For ease in writing, we do not show the size of a rank echelon matrix in the notation R_r; we rather specify it in different contexts. From Theorem 2.8, it follows that any matrix which is equivalent to R_r has also rank r.

Conversely, if a row of a matrix is a linear combination of other rows, then the rank of the matrix is same as that of the matrix obtained by deleting such a row. Similarly, deleting a column which is a linear combination of other columns does not change the rank of the matrix. It is thus possible to perform elementary row and column operations to bring a matrix of rank r to its rank echelon form R_r. We state and prove this result rigorously.

Theorem 3.12 (Rank factorization) *A matrix is of rank r iff it is equivalent to the rank echelon matrix R_r of the same size.*

Proof Let $A \in \mathbb{F}^{m \times n}$, and let R_r be the rank echelon matrix of size $m \times n$. If A is equivalent to R_r, then $\text{rank}(A) = \text{rank}(R_r) = r$ by Theorem 2.8.

For the converse, suppose $\text{rank}(A) = r$. Convert A to its RREF $C = E_1 A$, where E_1 is a suitable product of elementary matrices. Now, each non-pivotal column of C is a linear combination of its r pivotal columns.

Consider the matrix C^t. The pivotal columns of C are now the rows e_i^t in C^t. Each other row in C^t is a linear combination of these rows. Exchange the rows of C^t so that the first r rows of the new matrix are $e_1^t, \ldots, e_r^t$ in that order. That is, we have a matrix E_2, which is a product of elementary matrices of Type 1, such that $E_2 C^t$ has the first r rows as $e_1^t, \ldots, e_r^t$ and the next $m - r$ rows are linear combinations of these r rows.

Use suitable elementary row operations to zero-out the bottom $m - r$ rows. This is possible since each non-pivotal row is a linear combination of the pivotal rows $e_1^t, \ldots, e_r^t$. Thus, we obtain the matrix $E_3 E_2 C^t$, where E_3 is the suitable product of elementary matrices used in reducing the bottom $m - r$ rows to zero rows.

Then taking transpose, we see that $C E_3^t E_2^t$ is a matrix whose first r columns are $e_1, \ldots, e_r$ and all other columns are zero columns.

To summarize, we have obtained three matrices E_1, E_2, and E_3, which are products of elementary matrices such that

$$CE_3^t E_2^t = E_1 A E_3^t E_2^t = \begin{bmatrix} I_r & 0 \\ 0 & 0 \end{bmatrix} = R_r.$$

Since E_1, E_2^t and E_3^t are invertible, A and R_r are equivalent. ∎

Theorem 3.12 asserts that given any matrix $A \in \mathbb{F}^{m \times n}$ of rank r, there exist invertible matrices $P \in \mathbb{F}^{n \times n}$ and $Q \in \mathbb{F}^{m \times m}$ such that $A = Q^{-1} R_r P$. Thus, it is named as *rank factorization*.

The rank factorization can be used to characterize equivalence of matrices. If A and B are equivalent matrices, then clearly, they have the same rank. Conversely, if two $m \times n$ matrices have the same rank r, then both of them are equivalent to R_r. Then, they are equivalent. We thus obtain the *rank theorem*:

Two matrices of the same size are equivalent iff they have the same rank.

Observe that $R_r = \begin{bmatrix} I_r & 0 \\ 0 & 0 \end{bmatrix} = \begin{bmatrix} I_r \\ 0 \end{bmatrix} \begin{bmatrix} I_r & 0 \end{bmatrix}$. Due to rank factorization, there exist invertible matrices P and Q such that

$$A = Q^{-1} R_r P = Q^{-1} \begin{bmatrix} I_r \\ 0 \end{bmatrix} \begin{bmatrix} I_r & 0 \end{bmatrix} P = B C,$$

where

$$B = Q^{-1} \begin{bmatrix} I_r \\ 0 \end{bmatrix} \in \mathbb{F}^{m \times r}, \quad C = \begin{bmatrix} I_r & 0 \end{bmatrix} P \in \mathbb{F}^{r \times n}.$$

Here, rank$(B) = r = $ the number of columns in B. Similarly, rank$(C) = r = $ the number of rows in C. A matrix whose rank is equal to the number of rows or the number of columns is called a *full rank matrix*. The above factorization $A = BC$ shows that

each matrix can be written as a product of full rank matrices.

This result is known as the *full rank factorization*.

The notion of equivalence stems from the change of bases in both the domain and the co-domain of a matrix viewed as a linear transformation. In case the matrix is a square matrix of order n, it is considered as a linear transformation on $\mathbb{F}^{n \times 1}$. If we change the basis in $\mathbb{F}^{n \times 1}$, we would have a corresponding representation of the matrix in the new basis.

Let $A \in \mathbb{F}^{n \times n}$, a square matrix of order n. The matrix A is (viewed as) a linear transformation from $\mathbb{F}^{n \times 1}$ to $\mathbb{F}^{n \times 1}$. Let $E = \{e_1, \ldots, e_n\}$ be the standard basis of $\mathbb{F}^{n \times 1}$. The matrix A acts in the usual way: Ae_j is the jth column of A. Suppose we change the basis of $\mathbb{F}^{n \times 1}$ to $C = \{v_1, \ldots, v_n\}$. That is, in both the domain and the co-domain space, we take the new basis as C. If the equivalent matrix of A is M, then for each $v \in \mathbb{F}^{n \times 1}$,

$$[Av]_C = P^{-1}AP\,[v]_C, \quad P = \begin{bmatrix} v_1 & \cdots & v_n \end{bmatrix}.$$

That is, the coordinate matrix of A is $P^{-1}AP$. This leads to similarity of two matrices.

We say that two matrices $A, B \in \mathbb{F}^{n \times n}$ are **similar** iff $B = P^{-1}AP$ for some invertible matrix $P \in \mathbb{F}^{n \times n}$. Observe that in this case, B is the coordinate matrix of A with respect to the basis that comprises the columns of P.

Example 3.13 Consider the basis $N = \{[1, -1, 1]^t, [1, 1, -1]^t, [-1, 1, 1]^t\}$ for $\mathbb{R}^{3 \times 1}$. To determine the matrix similar to $A = \begin{bmatrix} 1 & 1 & 1 \\ -1 & 0 & 1 \\ 0 & 1 & 0 \end{bmatrix}$ when the basis has changed from the standard basis to N, we construct the matrix P by taking the basis vectors of N as in the following:

$$P = \begin{bmatrix} 1 & 1 & -1 \\ -1 & 1 & 1 \\ 1 & -1 & 1 \end{bmatrix}.$$

Then, the matrix similar to A when the basis changes to N is

$$B = P^{-1}AP = \tfrac{1}{2}\begin{bmatrix} 1 & 0 & 1 \\ 1 & 1 & 0 \\ 0 & 1 & 1 \end{bmatrix}\begin{bmatrix} 1 & 1 & 1 \\ -1 & 0 & 1 \\ 0 & 1 & 0 \end{bmatrix}\begin{bmatrix} 1 & 1 & -1 \\ -1 & 1 & 1 \\ 1 & -1 & 1 \end{bmatrix} = \tfrac{1}{2}\begin{bmatrix} 0 & 2 & 2 \\ 1 & -1 & 3 \\ -1 & -1 & 3 \end{bmatrix}.$$

From Example 3.12, we know that for $u = [1, 2, 3]^t$,

$$[u]_N = \begin{bmatrix} 2 \\ 3/2 \\ 5/2 \end{bmatrix}, \quad [Au]_N = \begin{bmatrix} 4 \\ 4 \\ 2 \end{bmatrix}, \quad B\,[u]_N = \tfrac{1}{2}\begin{bmatrix} 0 & 2 & 2 \\ 1 & -1 & 3 \\ -1 & -1 & 3 \end{bmatrix}\begin{bmatrix} 2 \\ 3/2 \\ 5/2 \end{bmatrix} = \begin{bmatrix} 4 \\ 4 \\ 2 \end{bmatrix}.$$

This verifies the condition $[Au]_N = B\,[u]_N$ for the vector $u = [1, 2, 3]^t$.   □

We emphasize that if $B = P^{-1}AP$ is a matrix similar to A, then the matrix A as a linear transformation on $\mathbb{F}^{n \times 1}$ with standard basis and the matrix B as a linear transformation on $\mathbb{F}^{n \times 1}$ with the basis consisting of columns of P are the same linear transformation. Moreover, if C is the basis whose jth element is the jth column of P, then for each vector $v \in \mathbb{F}^{n \times 1}$, $[Av]_C = P^{-1}AP\,[v]_C$.

Though equivalence is easy to characterize by the rank, similarity is much more difficult. And we postpone this to a later chapter.

Exercises for Sect. 3.7

1. Let $A \in \mathbb{C}^{m \times n}$. Define $T : \mathbb{C}^{1 \times m} \to \mathbb{C}^{1 \times n}$ by $T(x) = xA$ for $x \in \mathbb{C}^{1 \times m}$. Show that T is a linear transformation. Identify $T(e_j^t)$.
2. In each of the following cases, show that T is a linear transformation. Find the matrix A so that $T(x) = Ax$. Determine rank(A). And then, construct the full rank factorization of A.

(a) $T : \mathbb{R}^{3\times 1} \to \mathbb{R}^{2\times 1}$, $T([a, b, c]^t) = [c, b + a]^t$

(b) Let $T : \mathbb{R}^{3\times 1} \to \mathbb{R}^{3\times 1}$, $T([a, b, c]^t) = [a + b, 2a - b - c, a + b + c]^t$

3. Which matrices are equivalent to, and which are similar to
    (a) the zero matrix   (b) the identity matrix?

4. Consider the bases $O = \{u_1, u_2\}$ and $N = \{v_1, v_2\}$ for $\mathbb{F}^{2\times 1}$, where $u_1 = [1, 1]^t$, $u_2 = [-1, 1]^t$, $v_1 = [2, 1]^t$, and $v_2 = [1, 0]^t$. Let $A = \begin{bmatrix} -1 & 0 \\ 0 & 1 \end{bmatrix}$.

(a) Compute $Q = [A]_{O,O}$ and $R = [A]_{N,N}$.

(b) Find the change of basis matrix $P = I_{N,O}$.

(c) Compute $S = PQP^{-1}$.

(d) Is it true that $R = S$? Why?

(e) If $S = [s_{ij}]$, verify that $Av_1 = s_{11}v_1 + s_{21}v_2$, $Av_2 = s_{12}v_1 + s_{22}v_2$.

3.8 Problems

1. Let $A \in \mathbb{F}^{n\times n}$. Show that the following are equivalent to $\det(A) \neq 0$:

(a) The columns of A are linearly independent vectors in $\mathbb{F}^{n\times 1}$.

(b) The rows of A are linearly independent vectors in $\mathbb{F}^{1\times n}$.

2. Let $x, y, z \in \mathbb{F}^n$ be linearly independent. Determine whether the following are linearly independent:

(a) $x + y, y + z, z + x$ (b) $x - y, y - z, z - x$

3. Let U and V be subspaces of $\mathbb{F}^n$. Show the following:

(a) $U \cap V$ is a subspace of $\mathbb{F}^n$.

(b) $U \cup V$ need not be a subspace of $\mathbb{F}^n$.

(c) $U + V = \{u + v : u \in U, v \in V\}$ is a subspace of $\mathbb{F}^n$.

(d) If W is any subspace of $\mathbb{F}^n$, $U \subseteq W$, and $V \subseteq W$, then $U + V \subseteq W$.

(e) $\dim(U + V) + \dim(U \cap V) = \dim(U) + \dim(V)$.

4. Let $A \in \mathbb{F}^{m\times n}$. Show that if the columns of A are linearly independent, then $N(A) = \{0\}$. What if the rows of A are linearly independent?

5. Let $C = AB$, where $A \in \mathbb{F}^{m\times k}$ and $B \in \mathbb{F}^{k\times n}$. Show the following:

(a) If both A and B have linearly independent columns, then the columns of C are linearly independent.

(b) If both A and B have linearly independent rows, then the rows of C are linearly independent.

(c) If B has linearly dependent columns, then the columns of C are linearly dependent.

(d) If A has linearly dependent rows, then the rows of C are linearly dependent.

6. Let $A, B \in \mathbb{F}^{n\times n}$ be such that $N(A - B) = \mathbb{F}^{n\times 1}$. Show that $A = B$.

7. Let $A, B \in \mathbb{F}^{n \times n}$ be such that $AB = 0$. Show the following:
   (a) $R(B) \subseteq N(A)$ (b) $\text{rank}(A) + \text{rank}(B) \leq n$
8. Let $A = xy^t$ for nonzero vectors $x \in \mathbb{F}^{m \times 1}$ and $y \in \mathbb{F}^{n \times 1}$. Show that
   (a) $\text{span}\{x\} = R(A)$ (b) $\text{span}\{y^t\} = $ the row space of A
9. Is it true that if $A \in \mathbb{F}^{m \times n}$, then both A and A^t have the same nullity?
10. Let B be the RREF of an $m \times n$ matrix A. Are the following true?

    (a) The row spaces of A and B are equal.
    (b) The column spaces of A and B are equal.

11. Let U be a subspace of $\mathbb{F}^{n \times 1}$.

    (a) Does there exist a matrix in $\mathbb{F}^{n \times n}$ such that $U = R(A)$?
    (b) Does there exist a matrix in $\mathbb{F}^{n \times n}$ such that $U = N(A)$?

12. Let $A \in \mathbb{F}^{m \times n}$. Let $\{u_1, \ldots, u_k\}$ be a basis for $N(A)$. Argue that there exist vectors $v_1, \ldots, v_{n-k}$ so that $\{u_1, \ldots, u_k, v_1, \ldots, v_{n-k}\}$ is a basis for $\mathbb{F}^{n \times 1}$. Show that $\{Av_1, \ldots, Av_{n-k}\}$ is a basis for $R(A)$. This will give an alternate proof of the rank nullity theorem.
13. Let $A \in \mathbb{F}^{m \times n}$. Let $P \in \mathbb{F}^{n \times n}$ and $Q \in \mathbb{F}^{m \times m}$ be invertible matrices. Show the following:
    (a) $N(QA) = N(A)$ (b) $R(AP) = R(A)$
14. Using rank nullity theorem, deduce Theorem 2.8.
15. Let $A, B \in \mathbb{F}^{n \times n}$ have the same rank. Is it true that $\text{rank}(A^2) = \text{rank}(B^2)$?
16. Let $A, B \in \mathbb{F}^{n \times n}$. Show that A is similar to B iff there exist matrices $P, Q \in \mathbb{F}^{n \times n}$, with P invertible, such that $A = PQ$ and $B = QP$.
17. Let A and B be similar matrices. Are the following pairs similar?
    (a) A^t, B^t (b) A^{-1}, B^{-1} (c) A^k, B^k for $k \in \mathbb{N}$ (d) $A - \alpha I, B - \alpha I$
18. If A and B are similar matrices, show that $\text{tr}(A) = \text{tr}(B)$.
19. Let $A, B \in \mathbb{F}^{n \times n}$ have the same trace. Are A and B
    (a) similar?   (b) equivalent?
20. Let $A, B \in \mathbb{F}^{n \times n}$ have the same determinant. Are A and B
    (a) similar?   (b) equivalent?
21. Is a rank factorization of a matrix unique?
22. Is a full rank factorization of a matrix unique?
23. Using rank factorization, show that for any $m \times k$ matrix A and $k \times n$ matrix B, $\text{rank}(AB) \leq \min\{\text{rank}(A), \text{rank}(B)\}$.
24. Let A and B be $m \times n$ matrices. Using full rank factorization, show that $\text{rank}(A + B) \leq \text{rank}(A) + \text{rank}(B)$.
25. Using the rank theorem, prove that the row rank of a matrix is equal to its column rank.

Chapter 4
Orthogonality

4.1 Inner Products

The dot product in $\mathbb{R}^3$ is used to define the length of a vector and the angle between two nonzero vectors. In particular, the dot product is used to determine when two vectors become perpendicular to each other. This notion can be generalized to $\mathbb{F}^n$.

For vectors $u, v \in \mathbb{F}^{1 \times n}$, we define their **inner product** as

$$\langle u, v \rangle = uv^*.$$

In case, $\mathbb{F} = \mathbb{R}$, in the definition of inner product, x^* becomes x'. For example, if $u = [1, 2, 3]$, $v = [2, 1, 3]$, then $\langle u, v \rangle = 1 \times 2 + 2 \times 1 + 3 \times 3 = 13$.

Similarly, for $x, y \in \mathbb{F}^{n \times 1}$, we define their inner product as

$$\langle x, y \rangle = y^* x.$$

In numerical examples, if $u = (a_1, \ldots, a_n)$ and $v = (b_1, \ldots, b_n) \in \mathbb{F}^n$, then we may also write their inner product with the 'dot' notation. That is,

$$\langle u, v \rangle = (a_1, \ldots, a_n) \cdot (\bar{b}_1, \ldots, \bar{b}_n) = a_1 \bar{b}_1 + \cdots + a_n \bar{b}_n.$$

The inner product satisfies the following properties:

For all $x, y, z \in \mathbb{F}^n$ and for all $\alpha, \beta \in \mathbb{F}$,

1. $\langle x, x \rangle \geq 0$.
2. $\langle x, x \rangle = 0$ iff $x = 0$.
3. $\langle x, y \rangle = \overline{\langle y, x \rangle}$.
4. $\langle x + y, z \rangle = \langle x, z \rangle + \langle y, z \rangle$.
5. $\langle z, x + y \rangle = \langle z, x \rangle + \langle z, y \rangle$.
6. $\langle \alpha x, y \rangle = \alpha \langle x, y \rangle$.
7. $\langle x, \beta y \rangle = \bar{\beta} \langle x, y \rangle$.

In any vector space V, a function $\langle \ \cdot \ \rangle : V \times V \to \mathbb{F}$ that satisfies Properties (1)–(4) and (6) is called an *inner product*. And any vector space with an inner product is called an *inner product space*. Properties (5) and (7) follow from the others.

The inner product gives rise to the length of a vector as in the familiar case of $\mathbb{R}^{1\times 3}$. We now call the generalized version of length as the *norm*.

For $u \in \mathbb{F}^n$, we define its **norm**, denoted by $\|u\|$, as the non-negative square root of $\langle u, u \rangle$. That is,

$$\|u\| = \sqrt{\langle u, u \rangle}.$$

The norm satisfies the following properties:
For all $x, y \in \mathbb{F}^n$ and for all $\alpha \in \mathbb{F}$,

1. $\|x\| \geq 0$.
2. $\|x\| = 0$ iff $x = 0$.
3. $\|\alpha x\| = |\alpha| \, \|x\|$.
4. $|\langle x, y \rangle| \leq \|x\| \, \|y\|$. (*Cauchy-Schwartz inequality*)
5. $\|x + y\| = \|x\| + \|y\|$. (*Triangle inequality*)

A proof of Cauchy–Schwartz inequality goes as follows:
If $y = 0$, then both sides of the inequality are equal to 0. Else, $\langle y, y \rangle \neq 0$. Write $\alpha = \dfrac{\langle x, y \rangle}{\langle y, y \rangle}$. Then $\overline{\alpha}\langle x, y \rangle = \alpha\langle y, x \rangle = \alpha\,\overline{\alpha}\langle y, y \rangle = \dfrac{|\langle x, y \rangle|^2}{\|y\|^2}$. So,

$$0 \leq \langle x - \alpha y, x - \alpha y \rangle = \langle x, x \rangle - \overline{\alpha}\langle x, y \rangle - \alpha\langle y, x \rangle + \alpha\overline{\alpha}\langle x, y \rangle = \|x\|^2 - \frac{|\langle x, y \rangle|^2}{\|y\|^2}.$$

The triangle inequality is proved using Cauchy–Schwartz:

$$\begin{aligned}
\|x + y\|^2 &= \langle x + y, x + y \rangle = \|x\|^2 + \|y\|^2 + \langle x, y \rangle + \langle y, x \rangle \\
&= \|x\|^2 + \|y\|^2 + 2\,\mathrm{Re}\langle x, y \rangle \leq \|x\|^2 + \|y\|^2 + 2|\langle x, y \rangle| \\
&\leq \|x\|^2 + \|y\|^2 + 2\|x\| \, \|y\| = \big(\|x\| + \|y\|\big)^2.
\end{aligned}$$

Using these properties, the acute (non-obtuse) angle between any two nonzero vectors can be defined.

Let x and y be nonzero vectors in $\mathbb{F}^n$. The **angle between** x and y, denoted by $\theta(x, y)$, is defined by

$$\cos\theta(x, y) = \frac{|\langle x, y \rangle|}{\|x\| \, \|y\|}.$$

We single out a particular case. For vectors x, y we say that x is **orthogonal** to y, written as $x \perp y$, iff $\langle x, y \rangle = 0$.

Notice that this definition allows x and y to be zero vectors. Also, the zero vector is orthogonal to every vector. And, if $x \perp y$, then $y \perp x$; in this case, we say that x and y are orthogonal vectors.

It follows that if $x \perp y$, then $\|x\|^2 + \|y\|^2 = \|x + y\|^2$ for all vectors x, y. This is referred to as *Pythagoras law*. The converse of Pythagoras law holds in $\mathbb{R}^n$. That is, for all $x, y \in \mathbb{R}^n$, if $\|x + y\|^2 = \|x\|^2 + \|y\|^2$ then $x \perp y$. But it does not hold in $\mathbb{C}^n$, in general.

A set of nonzero vectors in $\mathbb{F}^n$ is called an **orthogonal set** in $\mathbb{F}^n$ iff each vector in the set is orthogonal to every other vector in the set. A singleton set with a nonzero vector in it is assumed to be orthogonal.

Theorem 4.1 *An orthogonal set of vectors is linearly independent.*

Proof Let $\{v_1, \ldots, v_m\}$ be an orthogonal set of vectors in $\mathbb{F}^n$. Then they are nonzero vectors. Assume that $\alpha_1 v_1 + \cdots + \alpha_m v_m = 0$. For $1 \le j \le m$,

$$\sum_{i=1}^{m} \alpha_i \langle v_i, v_j \rangle = \langle \alpha_1 v_1 + \cdots + \alpha_m v_m, v_j \rangle = \langle 0, v_j \rangle = 0.$$

If $i \ne j$, then $\langle v_i, v_j \rangle = 0$; and the sum above evaluates to $\alpha_j \langle v_j, v_j \rangle$. So, $\alpha_j \langle v_j, v_j \rangle = 0$. As $v_j \ne 0$, $\langle v_j, v_j \rangle \ne 0$. Hence $\alpha_j = 0$.

Therefore, $\{v_1, \ldots, v_m\}$ is linearly independent. ∎

A vector v with $\|v\| = 1$ is called a **unit vector**. An orthogonal set of unit vectors is called an **orthonormal set**. For instance, in $\mathbb{F}^3$,

$$\{(1, 2, 3), \ (2, -1, 0)\}$$

is an orthogonal set, but not orthonormal; whereas the following set is an orthonormal set:

$$\left\{ \left(\tfrac{1}{\sqrt{14}}, \tfrac{2}{\sqrt{14}}, \tfrac{3}{\sqrt{14}} \right), \ \left(\tfrac{2}{\sqrt{5}}, \tfrac{-1}{\sqrt{5}}, 0 \right) \right\}.$$

The standard basis $\{e_1, \ldots, e_n\}$ is an orthonormal set in $\mathbb{F}^n$.

Orthogonal and orthonormal sets enjoy nice properties; we will come across them in due course.

Exercises for Sect. 4.1

1. Compute $\langle (2, 1, 3), (1, 3, 2) \rangle$ and $\langle (1, 2i, 3), (1 + i, i, -3i) \rangle$.
2. Is $\{[1, 2, 3, -1]^t, \ [2, -1, 0, 0]^t, \ [0, 0, 1, 3]^t\}$ an orthogonal set in $\mathbb{F}^{4 \times 1}$?
3. Show that if $x, y \in \mathbb{F}^n$, then $x \perp y$ implies $\|x + y\|^2 = \|x\|^2 + \|y\|^2$.
4. Show that if $x, y \in \mathbb{R}^n$, then $\|x + y\|^2 = \|x\|^2 + \|y\|^2$ implies $x \perp y$.
5. In $\mathbb{C}$, the inner product is given by $\langle x, y \rangle = \overline{y}x$. Let $x = 1$ and $y = i$ be two vectors in $\mathbb{C}$. Show that $\|x\|^2 + \|y\|^2 = \|x + y\|^2$ but $\langle x, y \rangle \ne 0$.
6. Show that the *parallelogram law* holds in $\mathbb{F}^n$. That is, for all $x, y \in \mathbb{F}^n$, we have $\|x + y\|^2 + \|x - y\|^2 = 2(\|x\|^2 + \|y\|^2)$.
7. If x and y are orthogonal vectors in $\mathbb{F}^n$, then show that $\|x\| \le \|x + y\|$.
8. Construct an orthonormal set from $\{[1, 2, 0]^t, \ [2, -1, 0]^t, \ [0, 0, 2]^t\}$.

4.2 Gram–Schmidt Orthogonalization

An orthogonal set of vectors is linearly independent. Conversely, given n linearly independent vectors $v_1, \ldots, v_n$ (necessarily all nonzero), how do we construct an orthogonal set from these? Let us consider a particular case.

Consider two linearly independent vectors u_1, u_2 in $\mathbb{R}^3$. The span of $\{u_1, u_2\}$ is a plane passing through the origin. We would like to obtain two orthogonal vectors in the plane whose span is the same plane. Now, the set $\{u_1\}$ is as such orthogonal. To construct a vector orthogonal to u_1 in the plane, we draw a perpendicular from u_2 (from its end-point while its initial point is the origin) on u_1. Then, u_2 minus the foot of the perpendicular is expected to be orthogonal to u_2.

Suppose u_1 and u_2 have been drawn with their initial points at the origin O. Let A be the end-point of u_2. Draw a perpendicular from A to the straight line (extend if required) containing u_1. Suppose the foot of the perpendicular is B. Now, OBA is a right angled triangle. Let θ be the measure of the angle $\angle AOB$. Then

$$\frac{OB}{OA} = \cos\theta = \frac{\langle u_2, u_1 \rangle}{\|u_2\| \, \|u_1\|}.$$

The required vector $\overrightarrow{OB}$ is the length of OA times $\cos\theta$ times the unit vector in the direction of u_1. That is,

$$\overrightarrow{OB} = \|u_2\| \frac{\langle u_2, u_1 \rangle}{\|u_2\| \, \|u_1\|} \frac{u_1}{\|u_1\|} = \frac{\langle u_2, u_1 \rangle}{\langle u_1, u_1 \rangle} u_1.$$

Next, we define v_2 as the vector $u_2 - \overrightarrow{OB}$. Our construction says that

$$v_1 = u_1, \quad v_2 = u_2 - \frac{\langle u_2, v_1 \rangle}{\langle v_1, v_1 \rangle} v_1.$$

Clearly, $\operatorname{span}\{u_1, u_2\} = \operatorname{span}\{v_1, v_2\}$; and here is a verification of $v_2 \perp v_1$:

$$\langle v_2, v_1 \rangle = \left\langle u_2 - \frac{\langle u_2, v_1 \rangle}{\langle v_1, v_1 \rangle} v_1, \, v_1 \right\rangle = \langle u_2, v_1 \rangle - \frac{\langle u_2, v_1 \rangle}{\langle v_1, v_1 \rangle} \langle v_1, v_1 \rangle = 0.$$

If more than two linearly independent vectors in $\mathbb{F}^n$ are given, we may continue this process of taking away feet of the perpendiculars drawn from the last vector on all the previous ones, assuming that the previous ones have already been orthogonalized. It results in the process given in the following theorem.

Theorem 4.2 (Gram–Schmidt orthogonalization) *Let* $u_1, \ldots, u_m \in \mathbb{F}^n$. *Let* $1 \leq k \leq m$. *Define*

$$v_1 = u_1$$

$$v_2 = u_2 - \frac{\langle u_2, v_1 \rangle}{\langle v_1, v_1 \rangle} v_1$$

$$\vdots$$

$$v_k = u_k - \frac{\langle u_k, v_1 \rangle}{\langle v_1, v_1 \rangle} v_1 - \cdots - \frac{\langle u_k, v_{k-1} \rangle}{\langle v_{k-1}, v_{k-1} \rangle} v_{k-1}.$$

Then the following are true:

(1) $\text{span}\{v_1, \ldots, v_k\} = \text{span}\{u_1, \ldots, u_k\}$.
(2) If $\{u_1, \ldots, u_k\}$ *is linearly independent, then* $\{v_1, \ldots, v_k\}$ *is orthogonal.*
(3) If $u_k \in \text{span}\{u_1, \ldots, u_{k-1}\}$, *then* $v_k = 0$....,

Proof We use induction on k. For $k = 1$, $v_1 = u_1$. First, $\text{span}\{u_1\} = \text{span}\{v_1\}$. Second, if $\{u_1\}$ is linearly independent, then $u_1 \neq 0$. Thus, $\{v_1\} = \{u_1\}$ is orthogonal. Third, If $u_1 \in \text{span}\{u_1, \ldots, u_{k-1}\} = \text{span}(\varnothing) = \{0\}$, then $v_1 = u_1 = 0$.

Lay out the induction hypothesis that the statements in (1)–(3) are true for $k = j \geq 1$. We will show their truth for $k = j + 1$. For (1), we start with

$$v_{j+1} = u_{j+1} - \frac{\langle u_{j+1}, v_1 \rangle}{\langle v_1, v_1 \rangle} v_1 - \cdots - \frac{\langle u_{j+1}, v_j \rangle}{\langle v_j, v_j \rangle} v_j.$$

Clearly, $v_{j+1} \in \text{span}\{u_{j+1}, v_1, \ldots, v_j\}$ and $u_{j+1} \in \text{span}\{v_1, \ldots, v_{j+1}\}$. By the induction hypothesis, we have $\text{span}\{v_1, \ldots, v_j\} = \text{span}\{u_1, \ldots, u_j\}$. Thus, any vector which is a linear combination of $u_1, \ldots, u_j, u_{j+1}$ is a linear combination of $v_1, \ldots, v_j, u_{j+1}$; and then it is a linear combination of $v_1, \ldots, v_j, v_{j+1}$. Similarly, any vector which is a linear combination of $v_1, \ldots, v_j, v_{j+1}$ is also a linear combination of $u_1, \ldots, u_j, u_{j+1}$. Hence

$$\text{span}\{v_1, \ldots, v_j, v_{j+1}\} = \text{span}\{u_1, \ldots, u_j, u_{j+1}\}.$$

For (2), assume that $\{u_1, \ldots, u_j, u_{j+1}\}$ is linearly independent. Then as a subset of a linearly independent set, $\{u_1, \ldots, u_j\}$ is linearly independent. The induction hypothesis implies that $\{v_1, \ldots, v_j\}$ is orthogonal. For any i with $1 \leq i \leq j$, we have

$$\langle v_{j+1}, v_i \rangle = \langle u_{j+1}, v_i \rangle - \frac{\langle u_{j+1}, v_1 \rangle}{\langle v_1, v_1 \rangle} \langle v_1, v_i \rangle - \cdots - \frac{\langle u_{j+1}, v_j \rangle}{\langle v_j, v_j \rangle} \langle v_j, v_i \rangle.$$

Now, $\langle v_1, v_i \rangle = \cdots = \langle v_{i-1}, v_i \rangle = \langle v_{i+1}, v_i \rangle = \cdots = \langle v_j, v_i \rangle = 0$ due to the orthogonality of $\{v_1, \ldots, v_j\}$. Thus

$$\langle v_{j+1}, v_i \rangle = \langle u_{j+1}, v_i \rangle - \frac{\langle u_{j+1}, v_i \rangle}{\langle v_i, v_i \rangle} \langle v_i, v_i \rangle = 0.$$

Therefore, $\{v_1, \ldots, v_j, v_{j+1}\}$ is an orthogonal set.

For (3), suppose that $u_{j+1} \in \text{span}\{u_1, \ldots, u_j\}$. Since v_{j+1} is a linear combination of $u_{j+1}, v_1, \ldots, v_j$, and $\text{span}\{u_1, \ldots, u_j\} = \text{span}\{v_1, \ldots, v_j\}$, we have $v_{j+1} \in \text{span}\{v_1, \ldots, v_j\}$. That is, there exist scalars $\alpha_1, \ldots, \alpha_j$ such that

$$v_{j+1} = \alpha_1 v_1 + \cdots + \alpha_j v_j.$$

It implies $\langle v_{j+1}, v_{j+1} \rangle = \alpha_1 \langle v_1, v_{j+1} \rangle + \cdots + \alpha_j \langle v_j, v_{j+1} \rangle$. Due to the orthogonality of $\{v_1, \ldots, v_j, v_{j+1}\}$, we have $\langle v_{j+1}, v_1 \rangle = 0, \ldots, \langle v_{j+1}, v_j \rangle = 0$. So, $\langle v_{j+1}, v_{j+1} \rangle = 0$. This implies $v_{j+1} = 0$. ∎

Theorem 4.2 helps in constructing an orthogonal basis, determining linear independence, and also extracting a basis.

Given an ordered set (list) of vectors from $\mathbb{F}^n$, we apply Gram–Schmidt orthogonalization. If the zero vector has been obtained in the process, then the list is linearly dependent. Further, discarding all zero vectors, we obtain an orthogonal set, which is a basis for the span of the given vectors. Also, the vectors corresponding to the (nonzero) orthogonal vectors form a basis for the same subspace.

Example 4.1 Apply Gram–Schmidt Orthogonalization process on the set $\{u_1, u_2, u_3\}$, where $u_1 = (1, 0, 0, 0)$, $u_2 = (1, 1, 0, 0)$ and $u_3 = (1, 1, 1, 0)$.

$$\begin{aligned}
v_1 &= (1, 0, 0, 0). \\
v_2 &= u_2 - \frac{\langle u_2, v_1 \rangle}{\langle v_1, v_1 \rangle} v_1 = (1, 1, 0, 0) - 1(1, 0, 0, 0) = (0, 1, 0, 0). \\
v_3 &= u_3 - \frac{\langle u_3, v_1 \rangle}{\langle v_1, v_1 \rangle} v_1 - \frac{\langle u_3, v_2 \rangle}{\langle v_2, v_2 \rangle} v_2 \\
&= (1, 1, 1, 0) - (1, 0, 0, 0) - (0, 1, 0, 0) = (0, 0, 1, 0).
\end{aligned}$$

The vectors $(1, 0, 0, 0)$, $(0, 1, 0, 0)$, $(0, 0, 1, 0)$ are orthogonal; and they form a basis for the subspace $U = \text{span}\{u_1, u_2, u_3\}$ of $\mathbb{F}^4$. Also, $\{u_1, u_2, u_3\}$ is linearly independent, and it is a basis of U. □

Example 4.2 Apply Gram–Schmidt orthogonalization on $\{u_1, u_2, u_3\}$, where $u_1 = [1, 1, 0, 1]$, $u_2 = [0, 1, 1, -1]$ and $u_3 = [1, 3, 2, -1]$.

$$\begin{aligned}
v_1 &= [1, 1, 0, 1]. \\
v_2 &= u_2 - \frac{\langle u_2, v_1 \rangle}{\langle v_1, v_1 \rangle} v_1 = [0, 1, 1, -1] - 0[1, 1, 0, 1] = [0, 1, 1, -1]. \\
v_3 &= u_3 - \frac{\langle u_3, v_1 \rangle}{\langle v_1, v_1 \rangle} v_1 - \frac{\langle u_3, v_2 \rangle}{\langle v_2, v_2 \rangle} v_2
\end{aligned}$$

$$= [1, 3, 2, -1] - \frac{[1, 3, 2, -1] \cdot [1, 1, 0, 1]}{[1, 1, 0, 1] \cdot [1, 1, 0, 1]} [1, 1, 0, 1]$$

$$- \frac{[1, 3, 2, -1] \cdot [0, 1, 1, -1]}{[0, 1, 1, -1] \cdot [0, 1, 1, -1]} [0, 1, 1, -1]$$

$$= [1, 3, 2, -1] - [1, 1, 0, 1] - 2[0, 1, 1, -1] = [0, 0, 0, 0].$$

Since $v_3 = 0$, we have $U = \mathrm{span}\{u_1, u_2, u_3\} = \mathrm{span}\{v_1, v_2\}$. Further, discarding u_3 that corresponds to $v_3 = 0$, we see that $U = \mathrm{span}\{u_1, u_2\}$. Indeed, the process also revealed that $u_3 = v_1 - 2u_2 = u_1 - 2u_2$. Hence, $\{u_1, u_2\}$ and the orthogonal set $\{v_1, v_2\}$ are bases for U. □

Observe that in Gram–Schmidt process, at any stage if the computed vector turns out to be a zero vector, it is to be discarded, and the process carries over to the next stage.

Exercises for Sect. 4.2

1. Orthogonalize the vectors $(1, 1, 1)$, $(1, 0, 1)$, and $(0, 1, 2)$ using Gram–Schmidt process.
2. How do we use Gram–Schmidt to compute the rank of a matrix?
3. Construct orthonormal sets that are bases for $\mathrm{span}\{u_1, u_2, u_3\}$, where u_1, u_2, u_3 are the vectors in Examples 4.1–4.2.
4. Show that the cross product $u \times v$ of two linearly independent vectors u, v in $\mathbb{R}^{1 \times 3}$ is orthogonal to both u and v. How to obtain this third vector as $u \times v$ by Gram–Schmidt process?

4.3 QR-Factorization

An orthogonal (ordered) set can be made orthonormal by dividing each vector by its norm. Also you can modify Gram–Schmidt orthogonalization process to directly output orthonormal vectors. The modified version is as follows.

Gram–Schmidt Orthonormalization:

$$v_1 = u_1; \qquad\qquad\qquad\qquad w_1 = v_1/\|v_1\|$$
$$v_2 = u_2 - \langle u_2, w_1 \rangle w_1; \qquad\qquad w_2 = v_2/\|v_2\|$$
$$\vdots$$
$$v_m = u_m - \langle u_m, w_1 \rangle w_1 - \cdots - \langle u_m, w_{m-1} \rangle w_{m-1}; \quad w_m = v_m/\|v_m\|.$$

If during computation, we find that $v_k = 0$, then we take $w_k = 0$. We know that this can happen when $u_k \in \mathrm{span}\{u_1, \ldots, u_{k-1}\}$. Thus, linear dependence can also be determined in this orthonormalization process, as earlier. Further, such u_k, v_k and w_k may be discarded from the list for carrying out the process further in order to compute an orthonormal set that is a basis for the span of the vectors $u_1, \ldots, u_m$.

Given a list of linearly independent vectors $u_1, \ldots, u_m$, the orthonormalization process computes a list of orthonormal vectors $w_1, \ldots, w_m$ with the property that

$$\text{span}\{u_1, \ldots, u_k\} = \text{span}\{w_1, \ldots, w_k\} \quad \text{for } 1 \leq k \leq m.$$

Let V be a subspace of $\mathbb{F}^n$. An orthogonal (ordered) subset of V which is also a basis of V is called an **orthogonal basis** of V. Similarly, when an orthonormal (ordered) set is a basis of V, the set is said to be an **orthonormal basis** of V.

For example, the standard basis $\{e_1, \ldots, e_n\}$ of $\mathbb{F}^{n \times 1}$ is an orthonormal basis of $\mathbb{F}^{n \times 1}$. Similarly, $\{e_1^t, \ldots, e_n^t\}$ is an orthonormal basis of $\mathbb{F}^{1 \times n}$.

Since Gram–Schmidt processes construct orthogonal and orthonormal bases from a given basis of any subspace of $\mathbb{F}^n$, it follows that

every subspace of $\mathbb{F}^n$ has an orthogonal basis, and also an orthonormal basis.

Example 4.3 Applying Gram–Schmidt orthonormalization, construct an orthonormal basis for $U = \text{span}\{u_1, u_2, u_3, u_4\}$, where $u_1 = (1, 1, 0)$, $u_2 = (0, 1, 1)$, $u_3 = (1, 0, 1)$ and $u_4 = (2, 2, 2)$.

$$
\begin{aligned}
v_1 &= u_1 = (1, 1, 0) \\
w_1 &= v_1/\|v_1\| = \left(\tfrac{1}{\sqrt{2}}, \tfrac{1}{\sqrt{2}}, 0\right) \\
v_2 &= u_2 - \langle u_2, w_1 \rangle\, w_1 = (0, 1, 1) - \left((0, 1, 1) \cdot \left(\tfrac{1}{\sqrt{2}}, \tfrac{1}{\sqrt{2}}, 0\right)\right)\left(\tfrac{1}{\sqrt{2}}, \tfrac{1}{\sqrt{2}}, 0\right) \\
&= (0, 1, 1) - \tfrac{1}{\sqrt{2}}\left(\tfrac{1}{\sqrt{2}}, \tfrac{1}{\sqrt{2}}, 0\right) = \left(-\tfrac{1}{2}, \tfrac{1}{2}, 1\right) \\
w_2 &= v_2/\|v_2\| = \tfrac{\sqrt{2}}{\sqrt{3}}\left(-\tfrac{1}{2}, \tfrac{1}{2}, 1\right) = \left(-\tfrac{1}{\sqrt{6}}, \tfrac{1}{\sqrt{6}}, \tfrac{\sqrt{2}}{\sqrt{3}}\right) \\
v_3 &= u_3 - \langle u_3, w_1 \rangle\, w_1 - \langle u_3, w_2 \rangle\, w_2 \\
&= (1, 0, 1) - \left((1, 0, 1) \cdot \left(\tfrac{1}{\sqrt{2}}, \tfrac{1}{\sqrt{2}}, 0\right)\right)\left(\tfrac{1}{\sqrt{2}}, \tfrac{1}{\sqrt{2}}, 0\right) \\
&\quad - \left((1, 0, 1) \cdot \left(-\tfrac{1}{\sqrt{6}}, \tfrac{1}{\sqrt{6}}, \tfrac{\sqrt{2}}{\sqrt{3}}\right)\right)\left(-\tfrac{1}{\sqrt{6}}, \tfrac{1}{\sqrt{6}}, \tfrac{\sqrt{2}}{\sqrt{3}}\right) \\
&= (1, 0, 1) - \left(\tfrac{1}{2}, \tfrac{1}{2}, 0\right) - \left(-\tfrac{1}{6}, \tfrac{1}{6}, \tfrac{1}{3}\right) = \left(\tfrac{2}{3}, -\tfrac{2}{3}, \tfrac{2}{3}\right) \\
w_3 &= v_3/\|v_3\| = \tfrac{\sqrt{3}}{2}\left(\tfrac{2}{3}, -\tfrac{2}{3}, \tfrac{2}{3}\right) = \left(\tfrac{1}{\sqrt{3}}, -\tfrac{1}{\sqrt{3}}, \tfrac{1}{\sqrt{3}}\right) \\
v_4 &= u_4 - \langle u_4, w_1 \rangle\, w_1 - \langle u_4, w_2 \rangle\, w_2 - \langle u_4, w_3 \rangle\, w_3 = 0 \\
w_4 &= 0.
\end{aligned}
$$

The set $\left\{\left(\tfrac{1}{\sqrt{2}}, \tfrac{1}{\sqrt{2}}, 0\right),\ \left(-\tfrac{1}{\sqrt{6}}, \tfrac{1}{\sqrt{6}}, \tfrac{\sqrt{2}}{\sqrt{3}}\right),\ \left(\tfrac{1}{\sqrt{3}}, -\tfrac{1}{\sqrt{3}}, \tfrac{1}{\sqrt{3}}\right)\right\}$ is the required orthonormal basis of U. Notice that $U = \mathbb{F}^3$. $\square$

We have discussed two ways of extracting a basis for the span of a finite number of vectors from $\mathbb{F}^n$. One is the method of elementary row operations and the other is Gram–Schmidt orthogonalization or orthonormalization. The latter is a superior tool though computationally difficult. We now see one of its applications in factorizing a matrix.

Theorem 4.3 (QR-factorization) *Let the columns of $A \in \mathbb{F}^{m \times n}$ be linearly indepen-dent. Then there exist a matrix $Q \in \mathbb{F}^{m \times n}$ with orthonormal columns, and an upper triangular invertible matrix $R \in \mathbb{F}^{n \times n}$ such that $A = QR$. Further, if the diagonal entries in R are real, and they are either all positive, or all negative, then both Q and R are uniquely determined from A.*

Proof Write the columns of $A \in \mathbb{F}^{m \times n}$ as $u_1, u_2, \ldots, u_n$ in that order. These vectors are linearly independent in $\mathbb{F}^{m \times 1}$. So, $m \geq n$. Using Gram–Schmidt orthonormal-ization on the list of vectors $u_1, \ldots, u_n$, we obtain an orthonormal list of vectors, say, $w_1, w_2, \ldots, w_n$. Now, span$\{u_1, \ldots, u_k\}$ = span$\{w_1, \ldots, w_k\}$ for $1 \leq k \leq n$. In particular, u_k is a linear combination of $w_1, \ldots, w_k$.

Hence there exist scalars a_{ij}, $1 \leq i \leq j \leq n$ such that

$$
\begin{aligned}
u_1 &= a_{11}w_1 \\
u_2 &= a_{12}w_1 + a_{22}w_2 \\
&\ \vdots \\
u_k &= a_{1k}w_1 + \cdots + a_{kk}w_k \\
&\ \vdots \\
u_n &= a_{1n}w_1 + \cdots + a_{kn}w_k + \cdots + a_{nn}w_n
\end{aligned}
$$

We take $a_{ij} = 0$ for $i > j$, and form the matrix

$$
R = [a_{ij}] = \begin{bmatrix} a_{11} & a_{12} & \cdots & a_{1n} \\ & a_{22} & \cdots & a_{2n} \\ & & \ddots & \vdots \\ & & & a_{nn} \end{bmatrix}.
$$

Since $u_1, \ldots, u_k$ are linearly independent,

$$
u_k \notin \text{span}\{u_1, \ldots, u_{k-1}\} = \text{span}\{w_1, \ldots, w_{k-1}\}.
$$

Thus, a_{kk} is nonzero for each k. Then, R is an upper triangular invertible matrix.

Construct $Q = [w_1 \ \cdots \ w_n]$. Since $\{w_1, \ldots, w_n\}$ is an orthonormal set, $Q^*Q = I$. Of course, if all entries of A are real, then so are the entries of Q and R. In that case, $Q^t Q = I$. Moreover, for $1 \leq k \leq n$,

$$
\begin{aligned}
QRe_k &= Q[a_{1k}, \cdots, a_{kk}, 0, \cdots, 0]^t = Q(a_{1k}e_1 + \cdots + a_{kk}e_k) \\
&= a_{1k}Qe_1 + \cdots + a_{kk}Qe_k = a_{1k}w_1 + \cdots + a_{kk}w_k = u_k.
\end{aligned}
$$

That is, the kth column of QR is same as the kth column of A for each $k \in \{1, \ldots, n\}$. Therefore, $QR = A$.

For uniqueness of the factorization, suppose that

$$A = Q_1 R_1 = Q_2 R_2,$$

$Q_1, \ Q_2 \in \mathbb{F}^{m \times n}$ satisfy $Q_1^* Q_1 = Q_2^* Q_2 = I,$

$R_1 = [a_{ij}], \ R_2 = [b_{ij}] \in \mathbb{F}^{n \times n}$ are upper triangular, and

$a_{kk} > 0, \ b_{kk} > 0$ for each $k \in \{1, \ldots, n\}.$

We will see later what happens if the diagonal entries of R_1 and of R_2 are all negative. Then

$$R_1^* R_1 = R_1^* Q_1^* Q_1 R_1 = (Q_1 R_1)^* Q_1 R_1 = A^* A = (Q_2 R_2)^* Q_2 R_2 = R_2^* R_2.$$

Notice that $R_1, \ R_2, \ R_1^*$ and R_2^* are all invertible matrices. Multiplying $(R_2^*)^{-1}$ on the left, and $(R_1)^{-1}$ on the right, we have

$$(R_2^*)^{-1} R_1^* R_1 (R_1)^{-1} = (R_2^*)^{-1} R_2^* R_2 (R_1)^{-1}.$$

It implies

$$(R_2^*)^{-1} R_1^* = R_2 R_1^{-1}.$$

Here, the matrix on the left is a lower triangular matrix and that on the right is an upper triangular matrix. Therefore, both are diagonal. Comparing the diagonal entries in the products we have

$$[(b_{ii})^{-1}]^* a_{ii}^* = b_{ii} (a_{ii})^{-1} \quad \text{for } 1 \le i \le n.$$

That is, $|a_{ii}|^2 = |b_{ii}|^2$. Since $a_{ii} > 0$ and $b_{ii} > 0$, we see that $a_{ii} = b_{ii}$ for $1 \le i \le n$. Hence $(R_2^{-1})^* R_1^* = R_2 R_1^{-1} = I$. Therefore,

$$R_2 = R_1, \quad Q_2 = A R_2^{-1} = A R_1^{-1} = Q_1.$$

Observe that in the factorization $A = QR$, if all diagonal entries of R are negative, then we have $a_{kk} < 0$ and $b_{kk} < 0$ for all k. Now, $|a_{ii}|^2 = |b_{ii}|^2$ implies $a_{ii} = b_{ii}$, in the above proof. Consequently, uniqueness of Q and R follows. ∎

The uniqueness of QR-factorization states that if Q is selected from the set of all matrices with $Q^* Q = I$, and R is selected from the set of all upper triangular matrices with positive (or negative) diagonal entries, then there is only one such Q and there is only one such R that satisfy $A = QR$.

Observe that the columns of the matrix Q in the QR-factorization of a matrix A is computed by orthonormalizing the columns of A. And the the matrix R is computed by collecting the coefficients from the orthonormalization process. We may compute R by taking $R = Q^* A$.

Example 4.4 Consider the matrix $A = \begin{bmatrix} 1 & 1 \\ 0 & 1 \\ 1 & 1 \end{bmatrix}$ for QR-factorization.

We orthonormalize the columns of A to get Q, and then take $R = Q^t A$. So,

$$Q = \begin{bmatrix} 1/\sqrt{2} & 0 \\ 0 & 1 \\ 1/\sqrt{2} & 0 \end{bmatrix}, \quad R = Q^t A = \begin{bmatrix} \sqrt{2} & \sqrt{2} \\ 0 & 1 \end{bmatrix}.$$

It is easy to check that $A = QR$.

Further, if all diagonal entries of R are required to be positive, then the Q and R we have computed are the only possible matrices satisfying $A = QR$.

Observe that $A = (-Q)(-R)$ is also another QR-factorization of A, where the diagonal entries of R are negative. Again, this would be the only such QR-factorization of A. □

Exercises for Sect. 4.3

1. Find $u \in \mathbb{R}^{1 \times 3}$ so that $\{[1/\sqrt{3}, \ 1/\sqrt{3}, \ 1/\sqrt{3}], [1/\sqrt{2}, \ 0, \ -1/\sqrt{2}], \ u\}$ is an orthonormal set. Form a matrix with the vectors as rows, in that order. Verify that the columns of the matrix are also orthonormal.
2. Show that $\{(\frac{1+i}{2}, \frac{1-i}{2}), (\frac{i}{\sqrt{2}}, \frac{-1}{\sqrt{2}})\}$ is an orthonormal set in $\mathbb{C}^2$. Express $(2 + 4i, -2i)$ as a linear combination of the given orthonormal vectors.
3. Let $\{u, v\}$ be an orthonormal set in $\mathbb{C}^{2 \times 1}$. Let $x = (4 + i)u + (2 - 3i)v$. Determine the values of u^*x, v^*x, x^*u, x^*v, and $\|x\|$.
4. Let $\{u, v\}$ be an orthonormal basis for $\mathbb{R}^{2 \times 1}$. Let $x \in \mathbb{R}^{2 \times 1}$ be a unit vector. If $x^t u = \frac{1}{2}$, then what could be $x^t v$?
5. Let $A \in \mathbb{F}^{m \times n}$. Show the following:

   (a) For $m \leq n$, $AA^* = I$ iff the rows of A are orthonormal.
   (b) For $m \geq n$, $A^*A = I$ iff the columns of A are orthonormal.

6. Find a QR-factorization of each of the following matrices:

   (a) $\begin{bmatrix} 0 & 1 \\ 1 & 1 \\ 0 & 1 \end{bmatrix}$ (b) $\begin{bmatrix} 1 & 0 & 2 \\ 0 & 1 & 1 \\ 1 & 2 & 0 \end{bmatrix}$ (c) $\begin{bmatrix} 1 & 1 & 2 \\ 0 & 1 & -1 \\ 1 & 1 & 0 \\ 0 & 0 & 1 \end{bmatrix}$

4.4 Orthogonal Projection

Any vector $v = (a, b, c) \in \mathbb{F}^3$ can be expressed as $v = ae_1 + be_2 + ce_3$. Taking inner product of v with the basis vectors we find that $\langle v, e_1 \rangle = a$, $\langle v, e_2 \rangle = b$ and $\langle v, e_3 \rangle = c$. Then $v = \langle v, e_1 \rangle e_1 + \langle v, e_2 \rangle e_2 + \langle v, e_3 \rangle e_3$.

Such an equality holds for any orthonormal basis for a subspace of $\mathbb{F}^n$.

Theorem 4.4 *Let $\{v_1, \ldots, v_m\}$ be an orthonormal basis for a subspace V of $\mathbb{F}^n$. Then the following are true for each $x \in V$:*

(1) **(Fourier Expansion)** $x = \langle x, v_1 \rangle v_1 + \cdots + \langle x, v_m \rangle v_m$.
(2) **(Parseval Identity)** $\|x\|^2 = |\langle x, v_1 \rangle|^2 + \cdots + |\langle x, v_m \rangle|^2$.

Proof (1) Let $x \in V$. There exist scalars $\alpha_1, \ldots, \alpha_m$ such that

$$x = \alpha_1 v_1 + \cdots + \alpha_m v_m.$$

Let $i \in \{1, \ldots, m\}$. Taking inner product of x with v_i, and using the fact that $\langle v_i, v_j \rangle = 0$ for $j \neq i$, and $\langle v_i, v_i \rangle = 1$, we have

$$\langle x, v_i \rangle = \alpha_i \langle v_i, v_i \rangle = \alpha_i.$$

It then follows that $x = \langle x, v_1 \rangle v_1 + \cdots + \langle x, v_m \rangle v_m$.

(2) Let $x \in V$. By (1), $x = \sum_{j=1}^{m} \langle x, v_j \rangle v_j$. Then

$$\|x\|^2 = \langle x, x \rangle = \left\langle \sum_{j=1}^{m} \langle x, v_j \rangle v_j, \ x \right\rangle = \sum_{j=1}^{m} \langle x, v_j \rangle \langle v_j, x \rangle = \sum_{j=1}^{m} |\langle x, v_j \rangle|^2. \ \blacksquare$$

Recall that in Gram–Schmidt orthogonalization, we compute v_2 by taking away the foot of the perpendicular drawn from u_2 on u_1. For computation of v_{k+1}, we take away the feet of the perpendiculars drawn from u_{k+1} on the already constructed vectors $v_1, \ldots, v_k$. Let us use an abbreviation. For a subspace V of $\mathbb{F}^n$, and any vector $w \in \mathbb{F}^n$, we abbreviate

$$\text{for each } v \in V, \ w \perp v \quad \text{to} \quad w \perp V.$$

If $V = \text{span}\{v_1, \ldots, v_m\}$, then corresponding to any $v \in V$, we have scalars $\alpha_1, \ldots, \alpha_m$ such that $v = \alpha_1 v_1 + \cdots + \alpha_m v_m$. Now, if w is orthogonal to each v_i, then $w \perp v$. Hence, $w \perp V$ iff $w \perp v_i$ for each $i = 1, \ldots, m$.

Write $V_k = \text{span}\{v_1, \ldots, v_k\}$, and denote the foot of the perpendicular drawn from u_{k+1} on V_k by w. Gram–Schmidt orthogonalization process shows that $w \in V_k$ and $v_{k+1} - w \perp V_k$. It raises a question. Does the orthogonality condition $v_{k+1} - w \perp V_k$ uniquely determine a vector w from V_k?

Theorem 4.5 *Let V be a subspace of $\mathbb{F}^n$. Corresponding to each $x \in \mathbb{F}^n$, there exists a unique vector $y \in V$ such that $x - y \perp V$. Further, if $\{v_1, \ldots, v_m\}$ is any orthonormal basis of V, then such a vector y is given by $y = \sum_{j=1}^{m} \langle x, v_j \rangle v_j$.*

Proof The subspace V of $\mathbb{F}^n$ has an orthonormal basis. Let $\{v_1, \ldots, v_n\}$ be an orthonormal basis of V. Write

$$y = \langle x, v_1 \rangle v_1 + \cdots + \langle x, v_m \rangle v_m.$$

Now, $y \in V$. For $1 \leq i, j \leq m$, we have $\langle v_j, v_i \rangle = 0$ for $j \neq i$, and $\langle v_i, v_i \rangle = 1$. Then

$$\langle y, v_i \rangle = \left\langle \sum_{j=1}^{m} \langle x, v_j \rangle v_j, \ v_i \right\rangle = \sum_{j=1}^{m} \langle x, v_j \rangle \langle v_j, v_i \rangle = \langle x, v_i \rangle.$$

That is, $x - y \perp v_i$ for each $i \in \{1, \ldots, m\}$. It follows that $x - y \perp V$.

For uniqueness, let $y \in V$ and $z \in V$ be such that $\langle x - y, v \rangle = \langle x - z, v \rangle = 0$ for each $v \in V$. Then

$$\langle y - z, v \rangle = \langle (x - z) - (x - y), v \rangle = 0 \quad \text{for each } v \in V.$$

In particular, with $v = y - z$, we have $\langle y - z, y - z \rangle = 0$. It implies $y = z$. ∎

In view of Theorem 4.5, we give a name to such a vector y satisfying the orthogonality condition $x - y \perp V$. Intuitively, such a vector y is the foot of the perpendicular drawn from x to the subspace V. Notice that if $x \in V$, then the foot of the perpendicular drawn from x on V is x itself. We call this foot of the perpendicular as the **orthogonal projection of** x **onto the subspace** V, and write it as $\mathrm{proj}_V(x)$. That is,

Given $x \in \mathbb{F}^n$ and a subspace V of $\mathbb{F}^n$, $\mathrm{proj}_V(x)$ is the unique vector in V that satisfies the condition $x - \mathrm{proj}_V(x) \perp V$. Further, $\mathrm{proj}_V(x) = \sum_{j=1}^{m} \langle x, v_j \rangle v_j$, where $\{v_1, \ldots, v_m\}$ is any orthonormal basis of V.

Using this projection vector, we obtain the following useful inequality.

Theorem 4.6 (Bessel Inequality) *Let* $\{v_1, \ldots, v_m\}$ *be an orthonormal basis for a subspace* V *of* $\mathbb{F}^n$. *Then for each* $x \in \mathbb{F}^n$, $\|x\|^2 \geq \|\mathrm{proj}_V(x)\| = \sum_{j=1}^{m} |\langle x, v_j \rangle|^2$.

Proof Let $x \in \mathbb{F}^n$. Write $y = \mathrm{proj}_V(x) = \sum_{j=1}^{m} \langle x, v_j \rangle v_j$. Now, $y \in V$ and $x - y \perp v$ for each $v \in V$. In particular, $x - y \perp y$. Then Pythagoras theorem and Parseval identity imply that

$$\|x\|^2 = \|x - y\|^2 + \|y\|^2 \geq \|y\|^2 = \sum_{j=1}^{m} |\langle x, v_j \rangle|^2. \quad ∎$$

An alternate proof of Bessel inequality may be constructed by using extension of a basis of a subspace to the whole space. If $x \in \mathrm{span}\{v_1, \ldots, v_m\}$, then by Parseval identity, we have $\|x\|^2 = \sum_{j=1}^{m} |\langle x, v_j \rangle|^2$. Otherwise, we extend the orthonormal set $\{v_1, \ldots, v_m\}$ to a basis of $\mathbb{F}^n$; and then, use Gram–Schmidt orthonormalization to obtain an orthonormal basis $\{v_1, \ldots, v_m, v_{m+1}, \ldots, v_n\}$ for $\mathbb{F}^n$. Then Parseval identity gives $\|x\|^2 = \sum_{j=1}^{n} |\langle x, v_j \rangle|^2 \geq \sum_{j=1}^{m} |\langle x, v_j \rangle|^2$.

When working with column vectors, the projection vector $\mathrm{proj}_V(x)$ can be seen as a matrix product.

For this, let $\{v_1, \ldots, v_m\}$ be an orthonormal basis of a subspace V of $\mathbb{F}^{n \times 1}$. Let $x \in \mathbb{F}^{n \times 1}$. Write $z = [c_1, \cdots, c_m]^t \in \mathbb{F}^{m \times 1}$ with $c_j = \langle x, v_j \rangle = v_j^* x$, and $P = [v_1 \quad \cdots \quad v_m]$. Then

$$z = \begin{bmatrix} c_1 \\ \vdots \\ c_m \end{bmatrix} = \begin{bmatrix} v_1^* x \\ \vdots \\ v_m^* x \end{bmatrix} = P^* x, \quad y = \sum_{j=1}^{m} \langle x, v_j \rangle v_j = \sum_{j=1}^{m} c_j v_j = Pz = PP^* x.$$

Notice that $PP^* x = \mathrm{proj}_V(x)$ for each $x \in \mathbb{F}^{n \times 1}$. Due to this reason, the matrix $PP^* \in \mathbb{F}^{n \times n}$ is called the **projection matrix** that projects vectors of $\mathbb{F}^{n \times 1}$ onto the subspace V.

Example 4.5 Let $V = \text{span}\{[1, 0, -1]^t, \ [1, 1, 1]^t\}$ and let $x = [1, 2, 3]^t$. Compute the projection matrix that projects vectors of $\mathbb{F}^{3\times 1}$ onto V, and $\text{proj}_V(x)$.

Since the vectors $[1, 0, -1]^t$ and $[1, 1, 1]^t$ are orthogonal, an orthonormal basis for V is given by $\{v_1, v_2\}$, where

$$v_1 = [1/\sqrt{2}, \ 0, \ -1/\sqrt{2}]^t, \quad v_2 = [1/\sqrt{3}, \ 1/\sqrt{3}, \ 1/\sqrt{3}]^t.$$

With $P = [v_1 \ \ v_2]$, the projection matrix is given by

$$PP^* = \begin{bmatrix} 1/\sqrt{2} & 1/\sqrt{3} \\ 0 & 1/\sqrt{3} \\ -1/\sqrt{2} & 1/\sqrt{3} \end{bmatrix} \begin{bmatrix} 1/\sqrt{2} & 0 & -1/\sqrt{2} \\ 1/\sqrt{3} & 1/\sqrt{3} & 1/\sqrt{3} \end{bmatrix} = \begin{bmatrix} 5/6 & 1/3 & -1/6 \\ 1/3 & 1/3 & 1/3 \\ -1/6 & 1/3 & 5/6 \end{bmatrix}.$$

Then $\text{proj}_V(x) = PP^*x = [1, 2, 3]^t$.

Also, directly from the orthonormal basis for V, we obtain

$$\text{proj}_V(x) = \langle x, v_1 \rangle v_1 + \langle x, v_2 \rangle v_2 = [1, 2, 3]^t.$$

Indeed, $x = -1[1, 0, -1]^t + 2[1, 1, 1]^t \in V$. Thus, $\text{proj}_V(x) = x$. $\qquad\square$

Exercises for Sect. 4.4

1. Let $\{u, v, w\}$ be an orthonormal basis for a subspace of $\mathbb{R}^{5\times 1}$ and let $x = au + bv + cw$. If $\|x\| = 5$, $\langle x, u \rangle = 4$ and $x \perp v$, then what are the possible values of a, b and c?
2. Let A be the 4×2 matrix whose rows are $[1/2, \ -1/2], [1/2, \ 1/2], [1/2, \ -1/2]$ and $[1/2, \ 1/2]$.

   (a) Determine the projection matrix P that projects vectors onto $R(A)$.
   (b) Find an orthonormal basis for $N(A^t)$, and then determine the projection matrix Q that projects vectors onto $N(A^t)$.
   (c) Is it true that $P^2 = P^t = P$, and $Q^2 = Q^t = Q$?

3. Let y be the orthogonal projection of a vector $x \in \mathbb{F}^{n\times 1}$ onto a subspace V of $\mathbb{F}^{n\times 1}$. Show that $\|y\|^2 = \langle x, y \rangle$.

4.5 Best Approximation and Least Squares Solution

From the orthogonality condition, we guess that the length of $x - \text{proj}_V(x)$ is the smallest among the lengths of all vectors $x - v$, when v varies over V. If this intuition goes well, then $\text{proj}_V(x)$ would be closest to x compared to any other vector from V. Further, we may think of $\text{proj}_V(x)$ as an approximation of x from the subspace V. We show that our intuition is correct.

Let V be a subspace of $\mathbb{F}^n$ and let $x \in \mathbb{F}^n$. A vector $u \in V$ is called a **best approximation** of x from V iff $\|x - u\| \le \|x - v\|$ for each $v \in V$.

Theorem 4.7 *Let V be a subspace of $\mathbb{F}^n$ and let $x \in \mathbb{F}^n$. Then $\text{proj}_V(x)$ is the unique best approximation of x from V.*

Proof Let $y = \text{proj}_V(x)$. We show the following:

1. $\|x - y\|^2 \leq \|x - v\|^2$ for each $v \in V$.
2. If $u \in V$ satisfies $\|x - u\| \leq \|x - v\|$ for each $v \in V$, then $u = y$.

(1) Let $v \in V$. Since $y \in V$ and $y - v \in V$, by Theorem 4.5, $x - y \perp y - v$. By Pythagoras theorem,

$$\|x - v\|^2 = \|x - y + y - v\|^2 = \|x - y\|^2 + \|y - v\|^2 \geq \|x - y\|^2.$$

(2) Let $u \in V$ satisfy $\|x - u\| \leq \|x - v\|$ for each $v \in V$. In particular, with $v = y$, we have $\|x - u\| \leq \|x - y\|$. By (1), $\|x - y\| \leq \|x - u\|$. That is,

$$\|x - u\| = \|x - y\|.$$

Since $y - u \in V$, by Theorem 4.5, $x - y \perp y - u$. By Pythagoras theorem,

$$\|x - u\|^2 = \|x - y + y - u\|^2 = \|x - y\|^2 + \|y - u\|^2 = \|x - u\|^2 + \|y - u\|^2.$$

Hence $\|y - u\|^2 = 0$. Therefore, $u = y$. ∎

Given a subspace V of $\mathbb{F}^n$ and a vector $x \in \mathbb{F}^n$, Theorem 4.7 implies that a vector $y \in V$ is a best approximation of x from V iff the orthogonality condition is satisfied. Due to the uniqueness results, it provides two ways of computing the best approximation, which correspond to computing $\text{proj}_V(x)$ by employing an orthonormal basis, or by using the orthogonality condition directly.

Starting from any basis for V, we employ Gram–Schmidt orthonormalization to obtain an orthonormal basis $\{v_1, \ldots, v_m\}$ for V. Then the best approximation y of x from V is given by

$$y = \langle x, v_1 \rangle v_1 + \cdots + \langle x, v_m \rangle v_m.$$

Also, the best approximation may be computed by using the projection matrix.

In the second approach, we look for a vector y that satisfies the orthogonality condition:

$$x - y \perp v \quad \text{for each vector } v \in V.$$

If $\{v_1, \ldots, v_m\}$ is a basis for V, then this condition is equivalent to

$$x - y \perp v_j \quad \text{for each } j = 1, \ldots, m.$$

Since $y \in V$, we may write $y = \sum_{j=1}^{m} \alpha_j v_j$. Then we determine the scalars α_j so that for $i = 1, \ldots, m$, $\langle x - \sum_{j=1}^{m} \alpha_j v_j, v_i \rangle = 0$. That is, we solve the following linear system:

$$\langle v_1, v_1 \rangle \alpha_1 + \cdots + \langle v_m, v_1 \rangle \alpha_m = \langle x, v_1 \rangle$$
$$\vdots$$
$$\langle v_1, v_m \rangle \alpha_1 + \cdots + \langle v_m, v_m \rangle \alpha_m = \langle x, v_m \rangle$$

Theorem 4.7 guarantees that this linear system has a unique solution. Further, the system matrix of this linear system is $A = [a_{ij}]$, where $a_{ij} = \langle v_j, v_i \rangle$. Such a matrix which results by taking the inner products of basis vectors is called a *Gram matrix*. Theorem 4.7 implies that a Gram matrix is invertible. Can you prove directly that a Gram matrix is invertible?

Example 4.6 Find the best approximation of $x = (1, 0) \in \mathbb{R}^2$ from $V = \{(a, a) : a \in \mathbb{R}\}$.

An orthogonal basis for V is $\{(1/\sqrt{2}, 1/\sqrt{2})\}$. Thus, the best approximation of x from V is given by

$$\text{proj}_V(x) = \langle x, v_1 \rangle v_1 = ((1, 0) \cdot (1/\sqrt{2}, 1/\sqrt{2}))(1/\sqrt{2}, 1/\sqrt{2}) = (1/2, 1/2).$$

For illustration, we redo it using the projection matrix. Instead of $\mathbb{R}^2$, we now work in $\mathbb{R}^{2 \times 1}$. We have $x = [1, 0]^t$ and $V = \text{span}\{[1, 1]^t\}$. An orthonormal basis for V is $\{\frac{1}{\sqrt{2}}[1, 1]^t\}$. Then the projection matrix onto V is given by

$$PP^* = \frac{1}{\sqrt{2}}\begin{bmatrix} 1 \\ 1 \end{bmatrix} \frac{1}{\sqrt{2}} \begin{bmatrix} 1 & 1 \end{bmatrix} = \frac{1}{2}\begin{bmatrix} 1 & 1 \\ 1 & 1 \end{bmatrix}.$$

And the best approximation of x from V is $y = PP^*x = [1/2 \ \ 1/2]^t$.

In the second approach, we seek (α, α) so that $(1, 0) - (\alpha, \alpha) \perp (\beta, \beta)$ for all β. That is, to find (α, α) so that $(1 - \alpha, -\alpha) \cdot (1, 1) = 0$. So, $\alpha = 1/2$. The best approximation is $(1/2, 1/2)$ as obtained earlier. $\qquad\square$

We may use the technique of taking the best approximation for approximating a solution of a linear system.

Let $A \in \mathbb{F}^{m \times n}$ and let $b \in \mathbb{F}^{m \times 1}$. A vector $u \in \mathbb{F}^{n \times 1}$ is a called a **least squares solution** of the linear system $Ax = b$ iff $\|Au - b\| \leq \|Az - b\|$ for all $z \in \mathbb{F}^{m \times 1}$.

Theorem 4.7 implies that $u \in \mathbb{F}^{n \times 1}$ is a least squares solution of $Ax = b$ iff Au is the best approximation of b from $R(A)$. This best approximation Au can be computed uniquely from the orthogonality condition $Au - b \perp R(A)$. However, the vector u can be uniquely determined when A is one-one, that is, when the homogeneous system $Ax = 0$ has a unique solution. We summarize these facts in the following theorem.

Theorem 4.8 *Let $A \in \mathbb{F}^{m \times n}$ and let $b \in \mathbb{F}^{m \times 1}$.*

(1) The linear system $Ax = b$ has a least squares solution.
(2) A vector $u \in \mathbb{F}^{n \times 1}$ is a least squares solution of $Ax = b$ iff $Au - b \perp v$ for each $v \in R(A)$.
(3) A vector $u \in \mathbb{F}^{n \times 1}$ is a least squares solution of $Ax = b$ iff it is a solution of $A^*Ax = A^*b$.
(4) A least squares solution is unique iff $N(A) = \{0\}$.

Proof We prove only (3); others are obvious from the discussion we had. For this, let $u_1, \ldots, u_n$ be the n columns of A. Since the range space $R(A)$ is equal to $\text{span}\{u_1, \ldots, u_n\}$, by (2), we obtain

u is a least squares solution of $Ax = b$

iff $\langle Au - b, \, u_i \rangle = 0$, for $i = 1, \ldots, n$

iff $u_i^*(Au - b) = 0$ for $i = 1, \ldots, n$

iff $A^*(Au - b) = 0$

iff $A^* Au = A^* b$. ∎

If $A \in \mathbb{R}^{m \times n}$ and $b \in \mathbb{R}^{m \times 1}$, then a least squares solution u of the system $Ax = b$ satisfies $A^t Au = A^t b$.

Least squares solutions are helpful in those cases where some errors in data lead to an inconsistent system.

Example 4.7 Let $A = \begin{bmatrix} 1 & 1 \\ 0 & 0 \end{bmatrix}$, $b = \begin{bmatrix} 0 \\ 1 \end{bmatrix}$, and let $u = \begin{bmatrix} 1 \\ -1 \end{bmatrix}$.

We see that $A^t Au = A^t b$. Hence u is a least squares solution of $Ax = b$.
Notice that $Ax = b$ does not have a solution. □

Alternatively, we may use QR-factorization in computing a least squares solution of a linear system.

Theorem 4.9 *Let $A \in \mathbb{F}^{m \times n}$ have linearly independent columns. Let $A = QR$ be a QR-factorization of A. Then, the least squares solution of $Ax = b$ is given by $u = R^{-1} Q^* b$.*

Proof As columns of A are linearly independent, $\text{null}(A) = n - \text{rank}(A) = 0$. That is, $N(A) = \{0\}$. By Theorem 4.8, there exists a unique least squares solution u of $Ax = b$. Moreover, the least squares solution u of $Ax = b$ satisfies $A^* Au = A^* b$. Plugging in $A = QR$, we have $R^* Q^* QRu = R^* Q^* b$. As $Q^* Q = I$ and R^* is invertible, we obtain $Ru = Q^* b$. Thus, $u = R^{-1} Q^* b$. ∎

However, $u = R^{-1} Q^* b$ need not be a *solution of* $Ax = b$. The reason is, Q has orthonormal columns, but it need not have orthonormal rows. Consequently, QQ^* need not be equal to I. Then $Au = QRR^{-1}Q^* b = QQ^* b$ need not be equal to b.

On the other hand, if a solution v exists for $Ax = b$, then $Av = b$. It implies

$$0 = \|Av - b\| \leq \|Aw - b\| \quad \text{for each } w \in \mathbb{F}^{n \times 1}.$$

Therefore, every solution of $Ax = b$ is a least squares solution.

Observe that when a QR-factorization of A has already been obtained one computes the least squares solution by solving the linear system $Rx = Q^* b$, which is an easy task since R is upper triangular.

However, for such a computation of the least squares solution of $Ax = b$, the columns of A must be linearly independent. This implies that $m \geq n$. That is, the number of equations must not be less than the number of unknowns.

Exercises for Sect. 4.5

1. Find the best approximation of $x \in U$ from V where

(a) $U = \mathbb{R}^3$, $x = (1, 2, 1)$, $V = \operatorname{span}\{(3, 1, 2), (1, 0, 1)\}$
(b) $U = \mathbb{R}^3$, $x = (1, 2, 1)$, $V = \{(\alpha, \beta, \gamma) \in \mathbb{R}^3 : \alpha + \beta + \gamma = 0\}$
(c) $U = \mathbb{R}^4$, $x = (1, 0, -1, 1)$, $V = \operatorname{span}\{(1, 0, -1, 1), (0, 0, 1, 1)\}$

2. Find least squares solution(s) of the system $Ax = b$, where

(a) $A = \begin{bmatrix} 3 & 1 \\ 1 & 2 \\ 2 & -1 \end{bmatrix}$, $b = \begin{bmatrix} 1 \\ 0 \\ -2 \end{bmatrix}$ (b) $A = \begin{bmatrix} 1 & 1 & 1 \\ -1 & 0 & 1 \\ 1 & -1 & 0 \\ 0 & 1 & -1 \end{bmatrix}$, $b = \begin{bmatrix} 0 \\ 1 \\ -1 \\ -2 \end{bmatrix}$

4.6   Problems

1. For a nonzero vector v and any vector u, show that $\left\langle u - \dfrac{\langle u, v \rangle}{\|v\|^2} v, \, v \right\rangle = 0$. Then use Pythagoras theorem to derive Cauchy–Schwartz inequality.

2. Let $n > 1$. Using a unit vector $u \in \mathbb{R}^{n \times 1}$ construct infinitely many matrices $A \in \mathbb{R}^{n \times n}$ so that $A^2 = I$.

3. Let x and y be linearly independent vectors in $\mathbb{F}^{n \times 1}$. Let $A = xy^* + yx^*$. Show that $\operatorname{rank}(A) = 2$.

4. *Fundamental subspaces*: Let $A \in \mathbb{F}^{m \times n}$. Prove the following:

(a) $N(A) = \{x \in \mathbb{F}^{n \times 1} : x \perp y \text{ for each } y \in R(A^*)\}$.
(b) $N(A^*) = \{y \in \mathbb{F}^{m \times 1} : y \perp z \text{ for each } z \in R(A)\}$.
(c) $R(A) = \{y \in \mathbb{F}^{m \times 1} : y \perp z \text{ for each } z \in N(A^*)\}$.
(d) $R(A^*) = \{x \in \mathbb{F}^{n \times 1} : x \perp u \text{ for each } u \in N(A)\}$.

5. Let $A \in \mathbb{F}^{m \times n}$. Let B and E be bases for the subspaces $N(A)$ and $R(A^*)$ of $\mathbb{F}^{n \times 1}$, respectively. Show that $B \cup E$ is a basis of $\mathbb{F}^{n \times 1}$.

6. Find a basis for $N(A)$, where A has rows as $[1, 1, 1, -1]$ and $[1, 1, 3, 5]$. Using this basis, extend the orthonormal set $\left\{ \frac{1}{2}[1, 1, 1, -1]^t, \frac{1}{6}[1, 1, 3, 5]^t \right\}$ to an orthonormal set for $\mathbb{R}^{4 \times 1}$.

7. Let $A \in \mathbb{F}^{m \times n}$. Show that $N(A^*A) = N(A)$.

8. Let $\hat{x}$ be a least squares solution of the linear system $Ax = b$ for an $m \times n$ matrix A. Show that an $n \times 1$ vector y is a solution of $Ax = b$ iff $y = \hat{x} + z$ for some $z \in N(A)$.

9. Let $\{v_1, \dots, v_m\}$ be a basis for a subspace V of $\mathbb{F}^n$. Let $A = [a_{ij}]$ be the Gram matrix, where $a_{ij} = \langle v_j, v_i \rangle$. Show that the Gram matrix A satisfies $A^* = A$, and $x^*Ax > 0$ for each nonzero $x \in \mathbb{F}^m$. Conclude that the Gram matrix is invertible.

10. Let $A \in \mathbb{F}^{m \times n}$ have orthogonal columns and let $b \in \mathbb{F}^{m \times 1}$. Suppose that $y = [y_1, \cdots, y_n]^t$ is a least squares solution of $Ax = b$. Show that $\|A_{\star i}\|^2 y_i = b^* A_{\star i}$ for $i = 1, \dots, n$.

11. Let $A \in \mathbb{F}^{m \times n}$ have rank n. Let Q and R be the matrices obtained by Gram–Schmidt process applied on the columns of A, as in the QR-factorization. If $v = a_1 Q_{\star 1} + \cdots + a_n Q_{\star n}$ is the projection of a vector $b \in \mathbb{F}^{n \times 1}$ onto $R(A)$,

and $u = [a_1, \cdots, a_n]^*$, then show the following:

(a) $u = Q^*b$ (b) $v = QQ^*b$ (c) $QQ^* = A(A^*A)^{-1}A^*$

12. Let $A \in \mathbb{F}^{m \times n}$, P be the projection matrix that projects vectors of $\mathbb{F}^{m \times 1}$ onto $R(A)$, and let Q be the projection matrix that projects vector of $\mathbb{F}^{n \times 1}$ onto $R(A^*)$. Show the following:

   (a) $I - P$ is the projection matrix that projects vectors of $\mathbb{F}^{m \times 1}$ onto $N(A^*)$.

   (b) $I - Q$ is the projection matrix that projects vectors of $\mathbb{F}^{n \times 1}$ onto $N(A)$.

13. Let $A \in \mathbb{F}^{7 \times 5}$ have rank 4. Let P and Q be projection matrices that project vectors from $\mathbb{F}^{7 \times 1}$ onto $R(A)$ and $N(A^*)$, respectively. Show that $PQ = 0$ and $P + Q = I$.

14. Let $A \in \mathbb{R}^{n \times n}$ and let $b \in \mathbb{R}^{n \times 1}$ be a nonzero vector. Show that if b is orthogonal to each column of A, then the linear system $Ax = b$ is inconsistent. What are least squares solutions of $Ax = b$?

15. *Fredholm Alternative*: Let $A \in \mathbb{F}^{m \times n}$ and let $b \in \mathbb{F}^{m \times 1}$. Prove that exactly one of the following is true:

   (a) There exists a solution of $Ax = b$.

   (b) There exists $y \in \mathbb{F}^{m \times 1}$ such that $A^*y = 0$ and $y^*b \neq 0$.

Chapter 5
Eigenvalues and Eigenvectors

5.1 Invariant Line

Let $A = \begin{bmatrix} 0 & 1 \\ 1 & 0 \end{bmatrix}$. We view A as the linear transformation: $A : \mathbb{R}^{2 \times 1} \to \mathbb{R}^{2 \times 1}$. It transforms straight lines to straight lines or points. Does there exist a straight line which is transformed to itself? We see that

$$A \begin{bmatrix} x \\ y \end{bmatrix} = \begin{bmatrix} 0 & 1 \\ 1 & 0 \end{bmatrix} \begin{bmatrix} x \\ y \end{bmatrix} = \begin{bmatrix} y \\ x \end{bmatrix}.$$

Thus, the line $\{[x, x]^t : x \in \mathbb{R}\}$ never moves. So also the line $\{[x, -x]^t : x \in \mathbb{R}\}$. Observe that

$$A \begin{bmatrix} x \\ x \end{bmatrix} = 1 \begin{bmatrix} x \\ x \end{bmatrix}, \quad A \begin{bmatrix} x \\ -x \end{bmatrix} = (-1) \begin{bmatrix} x \\ -x \end{bmatrix}.$$

Let $A \in \mathbb{F}^{n \times n}$. A scalar $\lambda \in \mathbb{F}$ is called an **eigenvalue** of A iff there exists a nonzero vector $v \in \mathbb{F}^{n \times 1}$ such that $Av = \lambda v$. Such a nonzero vector v is called an **eigenvector of A for** (associated with, corresponding to) the eigenvalue λ.

Example 5.1 Let $A = \begin{bmatrix} 1 & 1 & 1 \\ 0 & 1 & 1 \\ 0 & 0 & 1 \end{bmatrix}$. We see that $A \begin{bmatrix} 1 \\ 0 \\ 0 \end{bmatrix} = 1 \begin{bmatrix} 1 \\ 0 \\ 0 \end{bmatrix}$. Therefore, A has an eigenvector $[1, 0, 0]^t$ associated with the eigenvalue 1. Is $[c, 0, 0]^t$ also an eigenvector associated with the same eigenvalue 1?

What are the other eigenvalues, and what are the corresponding eigenvectors of A? If $v = [a, b, c]^t \neq 0$ is an eigenvector for an eigenvalue λ, then the equation $Av = \lambda v$ implies that

$$a + b + c = \lambda a, \quad b + c = \lambda b, \quad c = \lambda c.$$

The last equation says that either $\lambda = 1$ or $c = 0$. If $c = 0$, then the second equation implies that either $\lambda = 1$ or $b = 0$. Then the first equation yields either $\lambda = 1$ or $a = 0$. Now, $c = 0$, $b = 0$, $a = 0$ is not possible, since it would lead to $v = 0$. In any case, $\lambda = 1$. Then the equations are simplified to

$$a + b + c = a, \quad b + c = b, \quad c = c.$$

It implies that $b = c = 0$. Then $v = [a, 0, 0]^t$ for any nonzero a. That is, such a vector v is an eigenvector for the only eigenvalue 1.   □

Example 5.2 Let $A = \begin{bmatrix} 1 & 1 & 1 \\ 2 & 2 & 2 \\ 3 & 3 & 3 \end{bmatrix}$. We see that $A \begin{bmatrix} a \\ b \\ c \end{bmatrix} = (a+b+c) \begin{bmatrix} 1 \\ 2 \\ 3 \end{bmatrix}$.

Therefore, when $a + b + c = 0$, we have 0 as an eigenvalue with eigenvector $[a, b, c]$. Verify that the vectors $[1, 0, -1]^t$, $[0, 1, -1]^t$ are eigenvectors for the eigenvalue 0.

For $a = 1$, $b = 2$, $c = 3$, we see that an eigenvalue is $a + b + c = 6$ with a corresponding eigenvector as $[1, 2, 3]^t$.

Does A have eigenvalues other than 0 and 6?   □

Corresponding to an eigenvalue, there are infinitely many eigenvectors.

Exercises for Sect. 5.1

1. Suppose $A \in \mathbb{F}^{n \times n}$, $\lambda \in \mathbb{F}$, and $b \in \mathbb{F}^{n \times 1}$ are such that $(A - \lambda I)x = b$ has a unique solution. Show that λ is not an eigenvalue of A.
2. Formulate a converse of the statement in Exercise 1 and prove it.
3. Let $A \in \mathbb{F}^{n \times n}$. Show that a scalar $\lambda \in \mathbb{F}$ is an eigenvalue of A iff the map $A - \lambda I : \mathbb{F}^{n \times 1} \to \mathbb{F}^{n \times 1}$ is not one-one.
4. Let $A \in \mathbb{F}^{n \times n}$. Show that A is invertible iff 0 is not an eigenvalue of A.
5. Let $A \in \mathbb{F}^{n \times n}$. If the entries in each column of A add up to a scalar λ, then find an eigenvector of A.

5.2   The Characteristic Polynomial

Eigenvalues of small matrices can be computed using the following theorem.

Theorem 5.1 *Let $A \in \mathbb{F}^{n \times n}$. A scalar $\lambda \in \mathbb{F}$ is an eigenvalue of A iff $\det(A - \lambda I) = 0$.*

Proof A scalar $\lambda \in \mathbb{F}$ is an eigenvalue of A
    iff $Av = \lambda v$ for some $v \neq 0$
    iff $(A - \lambda I)v = 0$ for some $v \neq 0$
    iff the linear system $(A - \lambda I)x = 0$ has a nonzero solution
    iff $\mathrm{rank}(A - \lambda I) < n$
    iff $\det(A - \lambda I) = 0$.   ■

The polynomial $(-1)^n \det(A - tI)$, which is also equal to $\det(tI - A)$, is called the **characteristic polynomial** of the matrix A; and it is denoted by $\chi_A(t)$. Each eigenvalue of A is a zero of the characteristic polynomial of A. Conversely, each zero of the characteristic polynomial is said to be a **complex eigenvalue** of A.

If A is a complex matrix of order n, then $\chi_A(t)$ is a polynomial of degree n in t. The *fundamental theorem of algebra* states that any polynomial of degree n, with complex coefficients, can be written as a product of n linear factors each with complex coefficients. Thus, $\chi_A(t)$ has exactly n, not necessarily distinct, zeros in $\mathbb{C}$. And these are the eigenvalues (complex eigenvalues) of the matrix A.

If A is a matrix with real entries, some of the zeros of $\chi_A(t)$ may turn out to be complex numbers with nonzero imaginary parts. Considering A as a linear transformation on $\mathbb{R}^{n \times 1}$, the scalars are now real numbers. Thus each zero of the characteristic polynomial may not be an eigenvalue; only the real zeros are.

However, if we regard a real matrix A as a matrix with complex entries, then A is a linear transformation on $\mathbb{C}^{n \times 1}$. Each complex eigenvalue, that is, a zero of the characteristic polynomial of A, is an eigenvalue of A.

Due to this obvious advantage, we consider a matrix in $\mathbb{R}^{n \times n}$ as one in $\mathbb{C}^{n \times n}$ so that each root of the characteristic polynomial of a matrix is considered an eigenvalue of the matrix. In this sense, an eigenvalue is taken as a complex eigenvalue, by default. Observe that when λ is an (a complex) eigenvalue of $A \in \mathbb{F}^{n \times n}$, a corresponding eigenvector x is a vector in $\mathbb{C}^{n \times 1}$, in general.

Example 5.3 Find the eigenvalues and corresponding eigenvectors of the matrix

$$A = \begin{bmatrix} 1 & 0 & 0 \\ 1 & 1 & 0 \\ 1 & 1 & 1 \end{bmatrix}.$$

Here, $\chi_A(t) = (-1)^3 \det(A - tI) = - \begin{vmatrix} 1-t & 0 & 0 \\ 1 & 1-t & 0 \\ 1 & 1 & 1-t \end{vmatrix} = (t-1)^3.$

The eigenvalues of A are its zeros, that is, $1, 1, 1$. To get an eigenvector, we solve $A[a, b, c]^t = [a, b, c]^t$ or that

$$a = a, \ a + b = b, \ a + b + c = c.$$

It gives $a = b = 0$, and $c \in \mathbb{F}$ can be arbitrary. Since an eigenvector is nonzero, all the eigenvectors are given by $(0, 0, c)^t$, for $c \neq 0$.   □

Example 5.4 For the matrix $A = \begin{bmatrix} 0 & 1 \\ -1 & 0 \end{bmatrix} \in \mathbb{R}^{2 \times 2}$, the characteristic polynomial is $\chi_A(t) = (-1)^2 \det(A - tI) = t^2 + 1$. It has no real zeros. Thus, A has no real eigenvalues. However, i and $-i$ are its complex eigenvalues. That is, the same matrix $A \in \mathbb{C}^{2 \times 2}$ has eigenvalues as i and $-i$. The corresponding eigenvectors are obtained by solving

$$A[a, b]^t = i[a, b]^t \quad \text{and} \quad A[a, b] = -i[a, b]^t.$$

For $\lambda = i$, we have $b = ia$, $-a = ib$. Thus, $[a, ia]^t$ is an eigenvector for $a \neq 0$. For the eigenvalue $-i$, the eigenvectors are $[a, -ia]$ for $a \neq 0$.

We consider A as a matrix with complex entries. With this convention, the matrix A has (complex) eigenvalues i and $-i$.   □

A polynomial with the coefficient of the highest degree as 1 is called a *monic polynomial*. In the characteristic polynomial, the factor $(-1)^n$ is multiplied with the determinant to make the result a monic polynomial. We see that

$$\chi_A(t) = (-1)^n \det(A - tI) = \det(tI - A) = t^n + a_{n-1}t^{n-1} + \cdots + a_1 t + a_0$$

for some scalars $a_0, \ldots, a_{n-1}$. If $\lambda_1, \ldots, \lambda_n$ are the complex eigenvalues of A, counting multiplicities, then

$$\chi_A(t) = (t - \lambda_1) \cdots (t - \lambda_n).$$

Exercises for Sect. 5.2

1. Find all eigenvalues and corresponding eigenvectors of the following matrices:

(a) $\begin{bmatrix} 3 & 0 & 0 & 0 \\ 0 & 2 & 0 & 0 \\ 0 & 0 & 0 & -2 \\ 0 & 0 & 2 & 0 \end{bmatrix}$ (b) $\begin{bmatrix} 1 & 2 & 3 & 4 & 5 \\ 1 & 2 & 3 & 4 & 5 \\ 1 & 2 & 3 & 4 & 5 \\ 1 & 2 & 3 & 4 & 5 \\ 1 & 2 & 3 & 4 & 5 \end{bmatrix}$

2. Find all eigenvalues and their corresponding eigenvectors of the $n \times n$ matrix A whose jth row has each entry j.
3. Let $p(t) = t^3 - \alpha t^2 - (\alpha + 3)t - 1$ for a real number α. Consider

$$A = \begin{bmatrix} \alpha & \alpha+3 & 1 \\ 1 & 0 & 0 \\ 0 & 1 & 0 \end{bmatrix}, \quad P = \begin{bmatrix} -1 & 2 & -\alpha-3 \\ 1 & -1 & \alpha+2 \\ -1 & 1 & -\alpha-1 \end{bmatrix}.$$

(a) Compute the characteristic polynomials of A and $P^{-1}AP$.
(b) Show that all zeros of $p(t)$ are real regardless of the value of α.

5.3 The Spectrum

If the characteristic polynomial of a 3×3 matrix A is $\chi_A(t) = (t - 1)^2(t - 2)$, then its eigenvalues are 1 and 2. Thus, the set of eigenvalues $\{1, 2\}$ does not give information as to how many times an eigenvalue is a repeated zero of the characteristic polynomial. In this case, the notion of a multiset comes of help. The sets $\{1, 1, 2\}$ and $\{1, 2, 2\}$ are both equal to the set $\{1, 2\}$, but the multiset $\{1, 1, 2\}$ is different from the multiset $\{1, 2, 2\}$. Again, the multiset $\{1, 1, 2\}$ is same as the multiset $\{1, 2, 1\}$. That is, the number of times an element occurs in a multiset is significant whereas the ordering of elements in a multiset is ignored.

The multiset of all complex eigenvalues of a matrix $A \in \mathbb{F}^{n \times n}$ is called the **spectrum** of A; and we denote it by $\sigma(A)$. If $A \in \mathbb{F}^{n \times n}$ has the (complex) eigenvalues $\lambda_1, \ldots, \lambda_n$, counting multiplicities, then the spectrum of A is $\sigma(A) = \{\lambda_1, \ldots, \lambda_n\}$; and vice versa. Notice that the spectrum of $A \in \mathbb{F}^{n \times n}$ always has n elements; though they may not be distinct. The following theorem lists some important facts about eigenvalues, using the notion of spectrum.

Theorem 5.2 *Let $A \in \mathbb{F}^{n \times n}$. Then the following are true.*

1. $\sigma(A^t) = \sigma(A)$.
2. *If B is similar to A, then $\sigma(B) = \sigma(A)$.*
3. *If $A = [a_{ij}]$ is a diagonal or an upper triangular or a lower triangular matrix, then $\sigma(A) = \{a_{11}, \ldots, a_{nn}\}$.*
4. *If $\sigma(A) = \{\lambda_1, \ldots, \lambda_n\}$, then $\det(A) = \lambda_1 \cdots \lambda_n$ and $\operatorname{tr}(A) = \lambda_1 + \cdots + \lambda_n$.*

Proof (1) $\chi_{A^t}(t) = \det(A^t - tI) = \det((A - tI)^t) = \det(A - tI) = \chi_A(t)$.
(2) $\chi_{P^{-1}AP}(t) = \det(P^{-1}AP - tI) = \det(P^{-1}(A - tI)P) = \det(P^{-1})\det(A - tI)$ $\det(P) = \det(A - tI) = \chi_A(t)$.
(3) In all these cases, $\chi_A(t) = \det(A - tI) = (a_{11} - t) \cdots (a_{nn} - t)$.
(4) Let $\sigma(A) = \{\lambda_1, \ldots, \lambda_n\}$. Then

$$\chi_A(t) = \det(A - tI) = (\lambda_1 - t) \cdots (\lambda_n - t). \tag{5.1}$$

Substituting $t = 0$, we obtain $\det(A) = \lambda_1 \cdots \lambda_n$.
For the other equality, we compute the coefficient of t^{n-1} using (5.1). Expanding $\det(A - tI)$ in its first row, we find that the first term in the expansion is $(a_{11} - t)\det(A_{11})$, where A_{11} is the minor corresponding to the $(1, 1)$th entry. All other terms are polynomials of degree less than or equal to $n - 2$.
Next, we expand $\det(A_{11})$; and continue in a similar way to obtain the following:
Coefficient of t^{n-1} in $\det(A - tI)$
$= $ Coefficient of t^{n-1} in $(a_{11} - t) \times \det(A_{11})$
$= \cdots = $ Coefficient of t^{n-1} in $(a_{11} - t) \times (a_{22} - t) \times \cdots \times (a_{nn} - t)$
$= (-1)^{n-1}(a_{11} + \cdots + a_{nn})$.
$= $ Coefficient of t^{n-1} in $(\lambda_1 - t) \cdots (\lambda_n - t)$, using (5.1)
$= (-1)^{n-1}(\lambda_1 + \cdots + \lambda_n)$.
Therefore, $\lambda_1 + \cdots + \lambda_n = a_{11} + \cdots + a_{nn} = \operatorname{tr}(A)$. ∎

Theorem 5.3 *Eigenvectors associated with distinct eigenvalues of a matrix are linearly independent.*

Proof Let $\lambda_1, \ldots, \lambda_k$ be the distinct eigenvalues of a matrix A, and let $v_1, \ldots, v_k$ be corresponding eigenvectors. We use induction on $j \in \{1, \ldots, k\}$.
For $j = 1$, since $v_1 \neq 0$, $\{v_1\}$ is linearly independent. Lay out the induction hypothesis (for $j = m$): the set $\{v_1, \ldots, v_m\}$ is linearly independent. To show that (for $j = m + 1$): the set $\{v_1, \ldots, v_m, v_{m+1}\}$ is linearly independent, let

$$\alpha_1 v_1 + \alpha_2 v_2 + \cdots + \alpha_m v_m + \alpha_{m+1} v_{m+1} = 0. \tag{5.2}$$

Then, $A(\alpha_1 v_1 + \alpha_2 v_2 + \cdots + \alpha_m v_m + \alpha_{m+1} v_{m+1}) = 0$ gives (since $A v_i = \lambda_i v_i$)

$$\alpha_1 \lambda_1 v_1 + \alpha_2 \lambda_2 v_2 + \cdots + \alpha_m \lambda_m v_m + \alpha_{m+1} \lambda_{m+1} v_{m+1} = 0.$$

Multiply (5.2) with λ_{m+1} and subtract from the last equation to get

$$\alpha_1 (\lambda_1 - \lambda_{m+1}) v_1 + \cdots + \alpha_m (\lambda_m - \lambda_{m+1}) v_m = 0.$$

By the Induction Hypothesis, $\alpha_i (\lambda_i - \lambda_{m+1}) = 0$ for $i = 1, 2, \ldots, m$. Since $\lambda_i \neq \lambda_{m+1}$, $\alpha_i = 0$ for each such i. Then, (5.2) yields $\alpha_{m+1} v_{m+1} = 0$. Then $v_{m+1} \neq 0$ implies $\alpha_{m+1} = 0$. This completes the proof of linear independence of eigenvectors associated with distinct eigenvalues. ∎

Exercises for Sect. 5.3

1. Let A be a 4×4 non-invertible matrix with all diagonal entries 1. If $2 + 3i$ is an eigenvalue of A, what are its other eigenvalues?
2. Show that if each eigenvalue of $A \in \mathbb{F}^{n \times n}$ has absolute value less than 1, then both $I - A$ and $I + A$ are invertible.
3. Show that if rank of an $n \times n$ matrix is 1, then its trace is one of its eigenvalues. What are its other eigenvalues?
4. Find the spectrum of a matrix where the sum of entries in each row is same.
5. Let $A, B, P \in \mathbb{C}^{n \times n}$ be such that $B = P^{-1} A P$. Let λ be an eigenvalue of A. Show that a vector v is an eigenvector of B corresponding to the eigenvalue λ iff $P v$ is an eigenvector of A corresponding to the same eigenvalue λ.
6. Do equivalent matrices have same eigenvalues?

5.4 Special Types of Matrices

A square matrix A is called **hermitian** iff $A^* = A$; **skew hermitian** iff $A^* = -A$; **symmetric** iff $A^t = A$; and **skew symmetric** iff $A^t = -A$.

Thus, a *real symmetric* matrix is real hermitian; and a *real skew symmetric* matrix is real skew hermitian. Hermitian matrices are also called *self-adjoint* matrices. In the following, B is symmetric, C is skew symmetric, D is hermitian, and E is skew hermitian; B is also hermitian and C is also skew hermitian.

$$B = \begin{bmatrix} 1 & 2 & 3 \\ 2 & 3 & 4 \\ 3 & 4 & 5 \end{bmatrix}, \ C = \begin{bmatrix} 0 & 2 & -3 \\ -2 & 0 & 4 \\ 3 & -4 & 0 \end{bmatrix}, \ D = \begin{bmatrix} 1 & -2i & 3 \\ 2i & 3 & 4 \\ 3 & 4 & 5 \end{bmatrix}, \ E = \begin{bmatrix} 0 & 2+i & 3 \\ 2-i & i & 4i \\ 3 & -4i & 0 \end{bmatrix}.$$

Notice that all diagonal entries of a hermitian matrix are real since $\bar{a}_{ii} = a_{ii}$. Similarly, each diagonal entry of a skew symmetric matrix must be zero, since $a_{ii} = -a_{ii}$. And each diagonal entry of a skew hermitian matrix must be 0 or purely imaginary, as $\bar{a}_{ii} = -a_{ii}$ implies $2\mathrm{Re}(a_{ii}) = 0$.

Let A be a square matrix. Since $A + A^t$ is symmetric and $A - A^t$ is skew symmetric, every square matrix can be written as a sum of a symmetric matrix and a skew symmetric matrix:

$$A = \tfrac{1}{2}(A + A^t) + \tfrac{1}{2}(A - A^t).$$

Similar rewriting is possible with hermitian and skew hermitian matrices:

$$A = \tfrac{1}{2}(A + A^*) + \tfrac{1}{2}(A - A^*).$$

A square matrix A is called **unitary** iff $A^*A = I = AA^*$. In addition, if A is real, then it is called an orthogonal matrix. That is, an **orthogonal matrix** is a matrix with real entries satisfying $A^tA = I = AA^t$.

Thus, a square matrix is unitary iff it is invertible and its inverse is equal to its adjoint. Similarly, a real matrix is orthogonal iff it is invertible and its inverse is its transpose. In the following, B is a unitary matrix of order 2, and C is an orthogonal matrix (also unitary) of order 3:

$$B = \frac{1}{2}\begin{bmatrix} 1+i & 1-i \\ 1-i & 1+i \end{bmatrix}, \quad C = \frac{1}{3}\begin{bmatrix} 2 & 1 & 2 \\ -2 & 2 & 1 \\ 1 & 2 & -2 \end{bmatrix}.$$

Example 5.5 For any $\theta \in \mathbb{R}$, the following are orthogonal matrices of order 2:

$$O_1 := \begin{bmatrix} \cos\theta & -\sin\theta \\ \sin\theta & \cos\theta \end{bmatrix}, \quad O_2 := \begin{bmatrix} \cos\theta & \sin\theta \\ \sin\theta & -\cos\theta \end{bmatrix}.$$

Let u be the vector in the plane that starts at the origin and ends at the point (a, b). Writing the point (a, b) as a column vector $[a, b]^t$, we see that

$$O_1\begin{bmatrix} a \\ b \end{bmatrix} = \begin{bmatrix} a\cos\theta - b\sin\theta \\ a\sin\theta + b\cos\theta \end{bmatrix}, \quad O_2\begin{bmatrix} a \\ b \end{bmatrix} = \begin{bmatrix} a\cos\theta + b\sin\theta \\ a\sin\theta - b\cos\theta \end{bmatrix}.$$

Thus, $O_1[a, b]^t$ is the end-point of the vector obtained by rotating the vector u by an angle θ. Similarly, $O_2[a, b]^t$ is a point obtained by reflecting u along a straight line that makes an angle $\theta/2$ with the x-axis. Accordingly, O_1 is called a *rotation* by an angle θ, and O_2 is called a *reflection* along a line making an angle of $\theta/2$ with the x-axis.

If $A = [a_{ij}]$ is an orthogonal matrix of order 2, then $A^tA = I$ implies

$$a_{11}^2 + a_{21}^2 = 1 = a_{12}^2 + a_{22}^2, \quad a_{11}a_{12} + a_{21}a_{22} = 0.$$

Then there exist $\alpha, \beta \in \mathbb{R}$ such that $a_{11} = \cos\alpha$, $a_{21} = \sin\alpha$, $a_{12} = \cos\beta$, $a_{22} = \sin\beta$ and $\cos(\alpha - \beta) = 0$. Thus $\alpha - \beta = \pm(\pi/2)$. It then follows that A is in either of the forms O_1 or O_2. That is, an orthogonal matrix of order 2 is either a rotation or a reflection.   □

Theorem 5.4 *Let A be a unitary or an orthogonal matrix of order n. Then the following are true:*

1. *For each pair of vectors $x, y \in \mathbb{F}^{n \times 1}$, $\langle Ax, Ay \rangle = \langle x, y \rangle$.*
2. *For each $x \in \mathbb{F}^{n \times 1}$, $\|Ax\| = \|x\|$.*
3. *The columns of A are orthonormal.*
4. *The rows of A are orthonormal.*
5. $|\det(A)| = 1$.

Proof (1) $\langle Ax, Ay \rangle = (Ay)^* Ax = y^* A^* Ax = y^* x = \langle x, y \rangle$.
(2) Take $x = y$ in (1).
(3) Let the columns of A be $u_1, \ldots, u_n$ in that order. Then $\langle u_j, u_i \rangle = u_i^* u_j$ is the (i, j)th entry of $A^* A$. Since $A^* A = I$, $\langle u_j, u_i \rangle = \delta_{ij}$. Therefore, $\{u_1, \ldots, u_n\}$ is an orthonormal set.
(4) Considering $AA^* = I$, we get this result.
(5) Notice that $\det(A^*) = \det(\overline{A}) = \overline{\det(A)}$. Thus

$$\det(A^* A) = \det(A^*)\det(A) = \overline{\det(A)}\det(A) = |\det(A)|^2.$$

However, $\det(A^* A) = \det(I) = 1$. Therefore, $|\det(A)| = 1$.   ∎

Theorem 5.4(1)–(2) show that unitary and orthogonal matrices preserve the inner product and the norm. Thus they are also called *isometries*. The condition $A^* A = I$ is equivalent to the condition that the columns of A are orthonormal. Similarly, the rows of A are orthonormal iff $AA^* = I$.

Theorem 5.4 implies that the determinant of an orthogonal matrix is either 1 or -1. It follows that the product of all eigenvalues, counting multiplicities, of an orthogonal matrix is ± 1.

The determinant of a hermitian matrix is a real number since $A = A^*$ implies

$$\det(A) = \det(A^*) = \det(\overline{A}) = \overline{\det(A)}.$$

We prove some interesting facts about the eigenvalues and eigenvectors of these special types of matrices.

Theorem 5.5 *Let $A \in \mathbb{F}^{n \times n}$ and let λ be any eigenvalue of A.*

1. *If A is hermitian, then $\lambda \in \mathbb{R}$; and eigenvectors corresponding to distinct eigenvalues of A are orthogonal.*
2. *If A is real symmetric, then $\lambda \in \mathbb{R}$; and there exists a real eigenvector corresponding to λ.*
3. *If A is skew hermitian or real skew symmetric, then λ is purely imaginary or zero.*
4. *If A is unitary or orthogonal, then $|\lambda| = 1$.*

Proof (1) Let A be hermitian; that is, $A^* = A$. Let $v \in \mathbb{C}^{n \times 1}$ be an eigenvector corresponding to the eigenvalue λ of A. Then

$$\lambda v^* v = v^*(\lambda v) = v^* A v = v^* A^* v = (Av)^* v = (\lambda v)^* v = \bar{\lambda} v^* v.$$

Since $v \neq 0$, we have $v^* v \neq 0$; consequently, $\bar{\lambda} = \lambda$. Therefore, $\lambda \in \mathbb{R}$.

For the second conclusion, let $\lambda \neq \mu$ be two distinct eigenvalues of A with corresponding eigenvectors as x and y. Both $\lambda \in \mathbb{R}$ and $\mu \in \mathbb{R}$. Then

$$\lambda y^* x = y^* \lambda x = y^* A x = y^* A^* x = (Ay)^* x = (\mu y)^* x = \bar{\mu} y^* x = \mu y^* x.$$

Hence $(\lambda - \mu) y^* x = 0$. As $\lambda \neq \mu$, we conclude that $y^* x = 0$. That is, $x \perp y$.

(2) Let A be real symmetric. Then $\lambda \in \mathbb{R}$ follows from (1). For the second statement, let $v = x + iy \in \mathbb{C}^{n \times 1}$ be an eigenvector of A corresponding to λ, with $x, y \in \mathbb{R}^{n \times 1}$. Then

$$A(x + iy) = \lambda(x + iy).$$

Comparing the real and imaginary parts, we have

$$Ax = \lambda x, \quad Ay = \lambda y.$$

Since $x + iy \neq 0$, at least one of x or y is nonzero. One such nonzero vector is a real eigenvector corresponding to the (real) eigenvalue λ of A.

(3) Let A be skew hermitian. That is, $A^* = -A$. Let $v \in \mathbb{C}^{n \times 1}$ be an eigenvector corresponding to the eigenvalue λ of A. Then

$$\lambda v^* v = v^*(\lambda v) = v^* A v = -v^* A^* v = -(Av)^* v = -(\lambda v)^* v = -\bar{\lambda} v^* v.$$

Since $v^* v \neq 0$, $\lambda = -\bar{\lambda}$. That is, $2\mathrm{Re}(\lambda) = 0$. This shows that λ is purely imaginary or zero.

When A is real skew symmetric, we take transpose instead of adjoint everywhere in the above proof.

(4) Suppose A is unitary, i.e., $A^* A = I$. Let v be an eigenvector corresponding to the eigenvalue λ. Now, $Av = \lambda v$ implies $v^* A^* = (\lambda v)^* = \bar{\lambda} v^*$. Then

$$v^* v = v^* I v = v^* A^* A v = \bar{\lambda} \lambda v^* v = |\lambda|^2 v^* v.$$

Since $v^* v \neq 0$, $|\lambda| = 1$.

Replacing A^* with A^t in the above proof yields the same conclusion when A is an orthogonal matrix. ∎

Existence of a real eigenvector corresponding to a hermitian matrix cannot be guaranteed, in general.

Example 5.6 Consider the matrix $A = \begin{bmatrix} 1 & 1-i \\ 1+i & -1 \end{bmatrix}$. It is a hermitian matrix. Its characteristic polynomial is $\chi_A(t) = t^2 - 3$. Its eigenvalues are $\pm\sqrt{3}$. To compute an eigenvector corresponding to the eigenvalue $\sqrt{3}$, we set up the linear system $(A - \sqrt{3}I)[a, \ b]^t = 0$. That is,

$$(1 - \sqrt{3})a + (1 - i)b = 0, \quad (1 + i)a - (1 + \sqrt{3})b = 0.$$

This equation has a nonzero solution $a = 1 - i$, $b = \sqrt{3} - 1$. So, the vector $v = [1 - i, \ \sqrt{3} - 1]^t$ is an eigenvector corresponding to the eigenvalue $\sqrt{3}$.

If both $a, b \in \mathbb{R}$, then comparing the imaginary parts in the equations, we get $-b = 0$ and $a = 0$. But that is impossible since $v \neq 0$. Hence, there does not exist a real eigenvector for the eigenvalue $\sqrt{3}$ of A.   □

Similarly, a real skew symmetric matrix can have purely imaginary eigenvalues. The only real eigenvalue of a real skew symmetric matrix is 0. Also, an orthogonal matrix can have complex eigenvalues so that each eigenvalue need not be ± 1. For instance, the matrix

$$\begin{bmatrix} 0 & -1 \\ 1 & 0 \end{bmatrix}$$

has eigenvalues $\pm i$. However, all real eigenvalues of an orthogonal matrix are ± 1; and any eigenvalue of an orthogonal matrix is of the form $e^{i\theta}$ for $\theta \in \mathbb{R}$.

Exercises for Sect. 5.4

1. Let $A \in \mathbb{F}^{n \times n}$. Show that for all $x, y \in \mathbb{F}^{n \times 1}$, $\langle Ax, Ay \rangle = \langle x, y \rangle$ iff for all $x \in \mathbb{F}^{n \times 1}$, $\|Ax\| = \|x\|$.
2. Construct a 3×3 hermitian matrix with no zero entries whose eigenvalues are 1, 2 and 3.
3. Construct a 2×2 real skew symmetric matrix whose eigenvalues are purely imaginary.
4. Show that if an invertible matrix is real symmetric, then so is its inverse.
5. Show that if an invertible matrix is hermitian, then so is its inverse.
6. Construct an orthogonal 2×2 matrix whose determinant is 1.
7. Construct an orthogonal 2×2 matrix whose determinant is -1.

5.5 Problems

1. Let $a + bi$ be an eigenvalue of $A \in \mathbb{R}^{n \times n}$ with an associated eigenvector $x + iy$, where $a, b \in \mathbb{R}$ and $x, y \in \mathbb{R}^{n \times 1}$. Show the following:

   (a) The vectors $x + iy$ and $x - iy$ are linearly independent in $\mathbb{C}^{n \times 1}$.
   (b) The vectors x and y are linearly independent in $\mathbb{R}^{n \times 1}$.

2. Let $A = [a_{ij}] \in \mathbb{R}^{2 \times 2}$, where $a_{12}a_{21} > 0$. Write $B = \operatorname{diag}(\sqrt{a_{21}/a_{12}}, 1)$. Let $C = B^{-1}AB$. What do you conclude about the eigenvalues and eigenvectors of C; and those of A?

3. Show that if λ is a nonzero eigenvalue of an $n \times n$ matrix A, then $N(A - \lambda I) \subseteq R(A)$.

4. An $n \times n$ matrix A is said to be *idempotent* if $A^2 = A$. Show that the only possible eigenvalues of an idempotent matrix are 0 or 1.

5. An $n \times n$ matrix A is said to be *nilpotent* if $A^k = 0$ for some natural number k. Show that 0 is the only eigenvalue of a nilpotent matrix.

6. Let $A \in \mathbb{F}^{n \times n}$ have an eigenvalue λ with an associated eigenvector x. Suppose A^* has an eigenvalue μ associated with y. If $\lambda \neq \bar{\mu}$, then show that $x \perp y$.

7. Let $A, B \in \mathbb{F}^{n \times n}$. Show that

$$\begin{bmatrix} I & A \\ 0 & I \end{bmatrix} \begin{bmatrix} 0 & 0 \\ B & BA \end{bmatrix} \begin{bmatrix} I & -A \\ 0 & I \end{bmatrix} = \begin{bmatrix} AB & 0 \\ B & 0 \end{bmatrix}.$$

Then deduce that $\chi_{AB}(t) = \chi_{BA}(t)$.

8. Let $a_0, a_1, \ldots, a_{n-1} \in \mathbb{F}$. Let $p(t) = t^n - a_{n-1}t^{n-1} - \cdots - a_0$ be a polynomial. Construct the matrix

$$C = \begin{bmatrix} a_{n-1} & a_{n-2} & \cdots & a_1 & a_0 \\ 1 & 0 & \cdots & 0 & 0 \\ 0 & 1 & \cdots & 0 & 0 \\ & & \ddots & & \\ 0 & 0 & \cdots & 1 & 0 \end{bmatrix}.$$

The matrix C is called the *companion matrix* of the polynomial $p(t)$. Show the following:

(a) $\chi_C(t) = p(t)$.

(b) If $p(\lambda) = 0$ for some $\lambda \in \mathbb{F}$, then λ is an eigenvalue of the matrix C with an associated eigenvector $[\lambda^{n-1}, \lambda^{n-2}, \cdots, \lambda, 1]^t$.

9. Let $A \in \mathbb{F}^{n \times n}$ and let $B = I - 2A + A^2$. Show the following:

(a) If 1 is an eigenvalue of A, then B is not invertible.

(b) If v is an eigenvector of A, then v is an eigenvector of B. Are the corresponding eigenvalues equal?

10. For which scalars α, are the $n \times n$ matrices A and $A + \alpha I$ similar?

11. Show that there do not exist $n \times n$ matrices A and B with $AB - BA = I$.

12. Let $A, B \in \mathbb{F}^{n \times n}$. Show that if 1 is not an eigenvalue of A, then the matrix equation $AX + B = X$ has a unique solution in $\mathbb{F}^{n \times n}$.

13. Let A and B be hermitian matrices. Show that AB is hermitian iff $AB = BA$.

14. Let A and B be hermitian matrices. Determine whether the matrices $A + B$, ABA, $AB + BA$, and $AB - BA$ are hermitian or not.

15. Let A be a non-hermitian matrix. Determine whether the matrices $A + A^*$, $A - A^*$, $A^*A - AA^*$, $(I + A)(A + A^*)$, and $(I + A)(I - A^*)$ are hermitian or not.

16. Show that a square matrix A is invertible iff $\det(A^*A) > 0$.

17. Let A be a skew hermitian matrix. Show that $I - A$ is invertible.

18. Let $A \in \mathbb{F}^{m \times n}$ have rank n. Let $P = A(A^*A)^{-1}A^*$. Show the following:

   (a) A^*A is invertible so that the construction of P is meaningful.
   (b) P is hermitian.
   (c) $Py = y$ for each $y \in R(A)$.
   (d) $P^k y = Py$ for each $y \in \mathbb{F}^{m \times 1}$ and for each $k \in \mathbb{N}$.
   (e) Let $y \in \mathbb{F}^{m \times 1}$ satisfy $y \perp Ax$ for each $x \in \mathbb{F}^{n \times 1}$. Then $Py = 0$.

19. Let $B \in \mathbb{F}^{n \times n}$. Suppose that $I + B$ is invertible. Let $C = (I + B)^{-1}(I - B)$. Show the following:

   (a) If B is skew hermitian, then C is unitary.
   (b) If B is unitary, then C is skew hermitian.

20. Show that in the plane,

   (a) a rotation following a rotation is a rotation.
   (b) a rotation following a reflection is a reflection.
   (c) a reflection following a rotation is a reflection.
   (d) a reflection following a reflection is a rotation.

21. Show that the eigenvalues of a reflection in the plane are ± 1. What could be the eigenvalues of a rotation?

22. Let $A \in \mathbb{R}^{2 \times 2}$ be an orthogonal matrix. Suppose that A has a non-trivial *fixed point*; that is, there exists a nonzero vector $v \in \mathbb{R}^{2 \times 1}$ such that $Av = v$. Show that with respect to any orthonormal basis B of $\mathbb{R}^{2 \times 1}$, the coordinate matrix $[A]_{B,B}$ is in the form
$$\begin{bmatrix} \cos \theta & \sin \theta \\ \sin \theta & -\cos \theta \end{bmatrix}.$$

23. Show that 1 is an eigenvalue of any 3×3 orthogonal matrix.

24. Using the previous problem, show that an orthogonal matrix of order 3 is similar to a matrix of the form $\begin{bmatrix} 1 & 0 \\ 0 & R \end{bmatrix}$, where R is either a reflection or a rotation in the plane.

25. Let A be an orthogonal matrix of order 3. Show that either $\sigma(A) = \{1, 1, 1\}$ or $\sigma(A) = \{1, e^{i\theta}, e^{-i\theta}\}$ for some $\theta \in \mathbb{R}$.

26. Show that the ith column of an orthogonal upper triangular matrix is either e_i or $-e_i$.

27. Let A be an orthogonal matrix with an eigenvalue 1 and a corresponding eigenvector v. Show that v is also an eigenvector of A^t.

28. Let $u_1, \ldots, u_n$ be the columns of an orthogonal matrix of order n. Show that
$$u_1 u_1^t + \cdots + u_n u_n^t = I.$$

29. Let $\{v_1, \ldots, v_n\}$ be an orthonormal set in $\mathbb{F}^{n \times 1}$. Let $\lambda_1, \ldots, \lambda_n \in \mathbb{R}$. Show that the matrix $A = \lambda_1 v_1 v_1^* + \cdots + \lambda_n v_n v_n^*$ is hermitian. Can you find eigenvalues and eigenvectors of A?

30. A *permutation matrix* of order n is an $n \times n$ matrix obtained from the identity matrix I_n by reordering its rows. Show the following:

    (a) A permutation matrix may be obtained by reordering the columns of the identity matrix.
    (b) A permutation matrix is a product of elementary matrices of the first type.
    (c) Let P be a permutation matrix of order m whose ith row is the k_ith row of I_m. Let A be an $m \times n$ matrix. Then PA is the $m \times n$ matrix whose ith row is the k_ith row of A.
    (d) Let Q be a permutation matrix of order n whose jth column is the k_jth column of I_n. Let A be an $m \times n$ matrix. Then AQ is the $m \times n$ matrix whose jth column is the k_jth column of A.
    (e) A permutation matrix is orthogonal.
    (f) If a permutation matrix P is symmetric, then $P^{2k} = I$ and $P^{2k+1} = P$ for $k = 0, 1, 2, \ldots$.

Chapter 6
Canonical Forms

6.1 Schur Triangularization

Eigenvalues and eigenvectors can be used to bring a matrix to nice forms using similarity transformations. A very general result in this direction is Schur's unitary triangularization. It says that using a suitable similarity transformation, we can represent a square matrix by an upper triangular matrix. Thus, the diagonal entries of the upper triangular matrix must be the eigenvalues of the given matrix.

Theorem 6.1 (Schur Triangularization) *Let $A \in \mathbb{C}^{n \times n}$. Then there exists a unitary matrix $P \in \mathbb{C}^{n \times n}$ such that $P^* A P$ is upper triangular. Moreover, if $A \in \mathbb{R}^{n \times n}$ and all eigenvalues of A are real, then P can be chosen to be an orthogonal matrix.*

Proof We use induction on n. If $n = 1$, then clearly A is an upper triangular matrix, and we take $P = [1]$, the identity matrix with a single entry as 1, which is both unitary and orthogonal.

Assume that for all $B \in \mathbb{C}^{m \times m}$, $m \geq 1$, we have a unitary matrix $Q \in \mathbb{C}^{m \times m}$ such that $Q^* B Q$ is upper triangular. Let $A \in \mathbb{C}^{(m+1) \times (m+1)}$. Let $\lambda \in \mathbb{C}$ be an eigenvalue of A with an associated eigenvector u. Recall that $\langle w, z \rangle = z^* w$ defines an inner product on $\mathbb{C}^{(m+1) \times 1}$. Then the unit vector $v = u / \|u\|$ is an eigenvector of A associated with the eigenvalue λ.

Extend the set $\{v\}$ to an orthonormal (ordered) basis $E = \{v, v_1, \ldots, v_m\}$ for $\mathbb{C}^{(m+1) \times 1}$. Here, you may have to use an extension of a basis, and then Gram–Schmidt orthonormalization process. Next, construct the matrix $R \in \mathbb{C}^{(m+1) \times (m+1)}$ by taking these basis vectors as its columns, in that order. That is, let

$$R = \begin{bmatrix} v & v_1 & \cdots & v_m \end{bmatrix}.$$

Since E is an orthonormal set, R is unitary. Consider the coordinate matrix of A with respect to the basis E. It is given by $[A]_{E,E} = R^{-1} A R = R^* A R$. The first column of $R^* A R$ is

© The Author(s), under exclusive license to Springer Nature Switzerland AG 2021
A. Singh, *Introduction to Matrix Theory*,
https://doi.org/10.1007/978-3-030-80481-7_6

$$R^*ARe_1 = R^*Av = R^{-1}\lambda v = \lambda R^{-1}v = \lambda R^{-1}Re_1 = \lambda e_1,$$

where $e_1 \in \mathbb{C}^{(m+1)\times 1}$ has first component as 1 and all other components 0. Then R^*AR can be written in the following block form:

$$R^*AR = \begin{bmatrix} \lambda & x \\ 0 & C \end{bmatrix},$$

where $0 \in \mathbb{C}^{m\times 1}$, $C \in \mathbb{C}^{m\times m}$, and $x = [v^*Av_1 \quad v^*Av_2 \quad \cdots \quad v^*Av_m] \in \mathbb{C}^{1\times m}$.

Notice that if $m = 1$, the proof is complete. For $m > 1$, by induction hypothesis, we have a matrix $S \in \mathbb{C}^{m\times m}$ such that S^*CS is upper triangular. Then take

$$P = R \begin{bmatrix} 1 & 0 \\ 0 & S \end{bmatrix}.$$

Since R and S are unitary, $P^*P = PP^* = I$;, that is, P is unitary. Moreover,

$$P^*AP = \begin{bmatrix} 1 & 0 \\ 0 & S \end{bmatrix}^* R^*AR \begin{bmatrix} 1 & 0 \\ 0 & S \end{bmatrix} = \begin{bmatrix} 1 & 0 \\ 0 & S^* \end{bmatrix} \begin{bmatrix} \lambda & x \\ 0 & C \end{bmatrix} \begin{bmatrix} 1 & 0 \\ 0 & S \end{bmatrix} = \begin{bmatrix} \lambda & xS \\ 0 & S^*CS \end{bmatrix}.$$

Of course, $xS \in \mathbb{C}^{1\times m}$. Since S^*CS is upper triangular, the induction proof is complete.

When $A \in \mathbb{R}^{n\times n}$, and all eigenvalues of A are real, we use the transpose instead of the adjoint everywhere in the above proof. Thus, P can be chosen to be an orthogonal matrix. ∎

To eradicate possible misunderstanding, we recall that A has only real eigenvalues means that when we consider this A as a matrix in $\mathbb{C}^{n\times n}$, all its complex eigenvalues turn out to be real numbers. This again means that all zeros of the characteristic polynomial of A are real.

We may unfold the inductive proof of Schur's triangularization as follows. Once we obtain a matrix similar to A in the form

$$\begin{bmatrix} \lambda & w \\ 0 & S^*CS \end{bmatrix},$$

we look for whether λ is still an eigenvalue of S^*CS. If so, we choose this eigenvalue over others for further reduction. In the next step, we obtain a matrix similar to A in the form

$$\begin{bmatrix} \lambda & w & y \\ 0 & \lambda & z \\ 0 & 0 & M \end{bmatrix}$$

where M is an $(n-2) \times (n-2)$ matrix. Continuing further this way, we see that a Schur triangularization of A exists, where on the diagonal of the final upper triangular matrix, equal eigenvalues occur together. This particular form will be helpful later. In general, the eigenvalues can occur on the diagonal of the Schur form in any prescribed order, depending on our choice of eigenvalue in each step.

Example 6.1 Consider the matrix $A = \begin{bmatrix} 2 & 1 & 0 \\ 2 & 3 & 0 \\ -1 & -1 & 1 \end{bmatrix}$ for Schur triangularization.

We find that $\chi_A(t) = (t-1)^2(t-4)$. All eigenvalues of A are real; thus, there exists an orthogonal matrix P such that $P^t A P$ is upper triangular. To determine such a matrix P, we take one of the eigenvalues, say 1. An associated eigenvector of norm 1 is $v = [0, 0, 1]^t$. We extend $\{v\}$ to an orthonormal basis for $\mathbb{C}^{3\times 1}$. For convenience, we take the orthonormal basis as

$$\{[0, 0, 1]^t, \ [1, 0, 0]^t, \ [0, 1, 0]^t\}.$$

Taking the basis vectors as columns, we form the matrix R as

$$R = \begin{bmatrix} 0 & 1 & 0 \\ 0 & 0 & 1 \\ 1 & 0 & 0 \end{bmatrix}.$$

We then find that

$$R^t A R = \begin{bmatrix} 1 & -1 & -1 \\ 0 & 2 & 1 \\ 0 & 2 & 3 \end{bmatrix}.$$

Next, we triangularize the matrix $C = \begin{bmatrix} 2 & 1 \\ 2 & 3 \end{bmatrix}$. It has eigenvalues 1 and 4. The eigenvector of norm 1 associated with the eigenvalue 1 is $[1/\sqrt{2}, \ -1/\sqrt{2}]^t$. We extend it to an orthonormal basis

$$\left\{[1/\sqrt{2}, \ -1/\sqrt{2}]^t, \ [1/\sqrt{2}, \ 1/\sqrt{2}]^t\right\}$$

for $\mathbb{C}^{2\times 1}$. Then we construct the matrix S by taking these basis vectors as its columns, that is,

$$S = \begin{bmatrix} 1/\sqrt{2} & 1/\sqrt{2} \\ -1/\sqrt{2} & 1/\sqrt{2} \end{bmatrix}.$$

We find that $S^t C S = \begin{bmatrix} 1 & -1 \\ 0 & 4 \end{bmatrix}$, which is an upper triangular matrix. Then

$$P = R \begin{bmatrix} 1 & 0 \\ 0 & S \end{bmatrix} = \begin{bmatrix} 0 & 1 & 0 \\ 0 & 0 & 1 \\ 1 & 0 & 0 \end{bmatrix} \begin{bmatrix} 1 & 0 & 0 \\ 0 & 1/\sqrt{2} & 1/\sqrt{2} \\ 0 & -1/\sqrt{2} & 1/\sqrt{2} \end{bmatrix} = \begin{bmatrix} 0 & 1/\sqrt{2} & 1/\sqrt{2} \\ 0 & -1/\sqrt{2} & 1/\sqrt{2} \\ 1 & 0 & 0 \end{bmatrix}.$$

$$\text{Consequently, } P^t A P = \begin{bmatrix} 1 & 0 & -\sqrt{2} \\ 0 & 1 & -1 \\ 0 & 0 & 4 \end{bmatrix}. \qquad \qquad \square$$

Since $P^* = P^{-1}$, Schur triangularization is informally stated as follows:

Every square matrix is unitarily similar to an upper triangular matrix.

Further, there is nothing sacred about being *upper* triangular. For, given a matrix $A \in \mathbb{C}^{n \times n}$, consider using Schur triangularization of A^*. There exists a unitary matrix Q such that $Q^* A^* Q$ is upper triangular. Then taking adjoint, we have $Q^* A Q$ is lower triangular. Thus, the following holds:

Every square matrix is unitarily similar to a lower triangular matrix.

Analogously, a real square matrix having only real eigenvalues is also orthogonally similar to a lower triangular matrix. We remark that the lower triangular form of a matrix need not be the transpose or the adjoint of its upper triangular form.

Moreover, neither the unitary matrix P nor the upper triangular matrix $P^* A P$ in Schur triangularization is unique. That is, there can be unitary matrices P and Q such that both $P^* A P$ and $Q^* A Q$ are upper triangular, and $P \neq Q$, $P^* A P \neq Q^* A Q$. The non-uniqueness stems from the choice of eigenvalues, their associated eigenvectors, and in extending those to an orthonormal basis. For instance, in Example 6.1, if you extend $\{[0, 0, 1]^t\}$ to the orthonormal basis $\{[0, 0, 1]^t, [0, 1, 0]^t, [1, 0, 0]^t\}$, then you end up with (Verify.)

$$P = \begin{bmatrix} 0 & -1/\sqrt{2} & 1/\sqrt{2} \\ 0 & 1/\sqrt{2} & 1/\sqrt{2} \\ 1 & 0 & 0 \end{bmatrix}, \quad P^t A P = \begin{bmatrix} 1 & 0 & -\sqrt{2} \\ 0 & 1 & 1 \\ 0 & 0 & 4 \end{bmatrix}.$$

Exercises for Sect. 6.1

1. Let $A = \begin{bmatrix} 1 & 2 \\ -1 & -2 \end{bmatrix}$, $B = \begin{bmatrix} -1 & 3 \\ 0 & 0 \end{bmatrix}$ and $C = \begin{bmatrix} 0 & 3 \\ 0 & -1 \end{bmatrix}$. Show that both B and C are Schur triangularizations of A.

2. Let $A = \begin{bmatrix} 1 & 2 \\ 3 & 6 \end{bmatrix}$, $B = \begin{bmatrix} 0 & 1 \\ 0 & 7 \end{bmatrix}$ and $C = \begin{bmatrix} 0 & 2 \\ 0 & 7 \end{bmatrix}$. Show that B is a Schur triangularization of A but C is not.

3. Determine a matrix A such that $A^* A = \begin{bmatrix} 4 & 0 & 0 \\ 0 & 1 & i \\ 0 & -i & 1 \end{bmatrix}$.

4. Using Schur triangularization prove the *Spectral mapping theorem*: for any polynomial $p(t)$, $\sigma(p(A)) = p(\sigma(A))$. In other words, for any polynomial $p(t)$, if λ is an eigenvalue of A, then $p(\lambda)$ is an eigenvalue of $p(A)$; and all (complex) eigenvalues of $p(A)$ are of this form.

6.2 Annihilating Polynomials

Schur triangularization brings a square matrix to an upper triangular form by a similarity transformation. For an upper triangular matrix $A \in \mathbb{F}^{n \times n}$ with diagonal entries $d_1, \ldots, d_n$, we wish to compute the product

$$(A - d_1 I)(A - d_2 I) \cdots (A - d_n I).$$

For example, let $A = \begin{bmatrix} d_1 & \star & \star \\ 0 & d_2 & \star \\ 0 & 0 & d_3 \end{bmatrix}$, where $\star$ stands for any entry possibly nonzero.

We see that

$$A - d_1 I = \begin{bmatrix} 0 & \star & \star \\ 0 & \star & \star \\ 0 & 0 & \star \end{bmatrix}$$

$$(A - d_1 I)(A - d_2 I) = \begin{bmatrix} 0 & \star & \star \\ 0 & \star & \star \\ 0 & 0 & \star \end{bmatrix} \begin{bmatrix} \star & \star & \star \\ 0 & 0 & \star \\ 0 & 0 & \star \end{bmatrix} = \begin{bmatrix} 0 & 0 & \star \\ 0 & 0 & \star \\ 0 & 0 & \star \end{bmatrix}$$

$$(A - d_1 I)(A - d_2 I)(A - d_3 I) = \begin{bmatrix} 0 & 0 & \star \\ 0 & 0 & \star \\ 0 & 0 & \star \end{bmatrix} \begin{bmatrix} \star & \star & \star \\ 0 & \star & \star \\ 0 & 0 & 0 \end{bmatrix} = \begin{bmatrix} 0 & 0 & 0 \\ 0 & 0 & 0 \\ 0 & 0 & 0 \end{bmatrix}.$$

What do you observe?

Theorem 6.2 *Let $A \in \mathbb{F}^{n \times n}$ be an upper triangular matrix with diagonal entries $d_1, \ldots, d_n$, in that order. Let $1 \leq k \leq n$. Then the first k columns of the product $(A - d_1 I) \cdots (A - d_k I)$ are zero columns.*

Proof For $k = 1$, we see that the first column of $A - d_1 I$ is a zero column. It means $(A - d_1 I)e_1 = 0$, where $e_1, \ldots, e_n$ are the standard basis vectors for $\mathbb{F}^{n \times 1}$. Assume that the result is true for $k = m < n$. That is,

$$(A - d_1 I) \cdots (A - d_m I)e_1 = 0, \ldots, (A - d_1 I) \cdots (A - d_m I)e_m = 0.$$

Notice that

$$(A - d_1 I) \cdots (A - d_m I)(A - d_{m+1} I) = (A - d_{m+1} I)(A - d_1 I) \cdots (A - d_m I).$$

It then follows that for $1 \leq j \leq m$,

$$(A - d_1 I) \cdots (A - d_m I)(A - d_{m+1} I)e_j$$
$$= (A - d_{m+1} I)(A - d_1 I) \cdots (A - d_m I)e_j = 0.$$

Next, $(A - d_{m+1}I)e_{m+1}$ is the $(m + 1)$th column of $A - d_{m+1}I$, which has all zero entries beyond the first m entries. That is, there are scalars $\alpha_1, \ldots, \alpha_m$ such that $(A - d_{m+1}I)e_{m+1} = \alpha_1 e_1 + \cdots + \alpha_m e_m$. Then

$$(A - d_1 I) \cdots (A - d_m I)(A - d_{m+1}I)e_{m+1}$$
$$= (A - d_1 I) \cdots (A - d_m I)(\alpha_1 e_1 + \cdots + \alpha_m e_m) = 0.$$

Therefore, the first $m + 1$ columns of $(A - d_1 I) \cdots (A - d_m I)(A - d_{m+1}I)$ are zero columns. By induction, the proof is complete. ∎

In what follows, we will be substituting the variable t by a square matrix in a polynomial $q(t)$. In such a substitution, we will interpret the constant term a_0 as $a_0 I$. For instance,

$$\text{if } A \in \mathbb{F}^{n \times n} \text{ and } q(t) = 2 + t + 5t^2, \text{ then } q(A) = 2I_n + A + 5A^2.$$

If $q(A)$ turns out to be the zero matrix, then we say that the polynomial $q(t)$ **annihilates** the matrix A; we also say that $q(t)$ is an **annihilating polynomial** of A; and that A is **annihilated by** $q(t)$. The zero polynomial annihilates every matrix. Does every matrix have an annihilating nonzero polynomial?

Theorem 6.3 (Cayley–Hamilton) *Each square matrix is annihilated by its characteristic polynomial.*

Proof Let $A \in \mathbb{F}^{n \times n}$. We consider A as a matrix in $\mathbb{C}^{n \times n}$. By Schur triangularization, there exist a unitary matrix $P \in \mathbb{C}^{n \times n}$ and an upper triangular matrix $U \in \mathbb{C}^{n \times n}$ such that $P^* A P = U$. That is, $A = P U P^*$. Then, for all scalars α and β, we have

$$\alpha I + \beta A = \alpha I + \beta P U P^* = P(\alpha I)P^* + P(\beta U)P^* = P(\alpha I + \beta U)P^*.$$

And $A^2 = P U P^* P U P^* = P U^2 P^*$. By induction, it follows that for any polynomial $q(t)$, $q(A) = P q(U) P^*$. In particular,

$$\chi_A(A) = P \chi_A(U) P^*.$$

Suppose the diagonal entries of U are $\lambda_1, \ldots, \lambda_n$. Since A and U are similar matrices,
$$\chi_A(t) = \chi_U(t) = (t - \lambda_1) \cdots (t - \lambda_n).$$

Theorem 6.2 implies that the n columns of $(U - \lambda_1 I) \cdots (U - \lambda_n I)$ are zero columns. That is, $\chi_A(U) = 0$. Therefore, $\chi_A(A) = P \chi_A(U) P^* = 0$. ∎

Cayley–Hamilton theorem helps us in computing powers of matrices and also the inverse of a matrix if it exists. Suppose $A \in \mathbb{F}^{n \times n}$ has the characteristic polynomial

$$\chi_A(t) = t^n + a_{n-1}t^{n-1} + \cdots + a_1 t + a_0.$$

By Cayley–Hamilton theorem, $A^n + a_{n-1}A^{n-1} + \cdots + a_1 A + a_0 I = 0$. Then

$$A^n = -(a_0 I + a_1 A + \cdots + a_{n-1}A^{n-1}).$$

Thereby, computation of A^n, A^{n+1}, ... can be reduced to that of A, ..., A^{n-1}.

Next, suppose that A is invertible. Then $\det(A) \neq 0$. Since $\det(A)$ is the product of all eigenvalues of A, $\lambda = 0$ is not an eigenvalue of A. It implies that t, which is equal to $(t - \lambda)$ for $\lambda = 0$, is not a factor of the characteristic polynomial of A. Therefore, the constant term a_0 in the characteristic polynomial of A is nonzero. By Cayley–Hamilton theorem,

$$a_0 I = -(a_1 A + a_2 A^2 + \cdots + a_{n-1}A^{n-1} + A^n).$$

Multiplying A^{-1} and simplifying, we obtain

$$A^{-1} = -\frac{1}{a_0}(a_1 I + a_2 A + \cdots + a_{n-1}A^{n-2} + A^{n-1}).$$

This way, A^{-1} can also be computed from A, A^2, ..., A^{n-1}.

Example 6.2 Let $A = \begin{bmatrix} 1 & 0 & 1 \\ 0 & 1 & 1 \\ 0 & 0 & 1 \end{bmatrix}$. Its characteristic polynomial is $\chi_A(t) = (t - 1)^3$.

Cayley–Hamilton theorem says that $(A - I)^3 = 0$. By direct computation, we find that

$$(A - I)^2 = \begin{bmatrix} 0 & 0 & 1 \\ 0 & 0 & 1 \\ 0 & 0 & 0 \end{bmatrix}^2 = \begin{bmatrix} 0 & 0 & 0 \\ 0 & 0 & 0 \\ 0 & 0 & 0 \end{bmatrix}.$$

Thus, it is not necessary that the degree of an annihilating polynomial of A has to be the order of A. □

Let $A \in \mathbb{F}^{n \times n}$. A monic polynomial of least degree that annihilates A is called the **minimal polynomial** of A.

In Example 6.2, no polynomial of degree 1 annihilates A. But the monic polynomial $(t - 1)^2$ annihilates it. Therefore, $(t - 1)^2$ is the minimal polynomial of A.

Observe that the degree of a minimal polynomial is unique. Moreover, if $q(t)$ is a minimal polynomial of A, and $q(t)$ has degree k, then no monic polynomial of degree less than k annihilates A. Therefore, no nonzero polynomial of degree less than k annihilates A. In other words,

if a polynomial of degree less than the degree of the minimal polynomial of A annihilates A, then this polynomial must be the zero polynomial.

Cayley–Hamilton theorem implies that the degree of the minimal polynomial does not exceed the order of the matrix. However, any multiple of an annihilating polynomial also annihilates the given matrix. For instance, the polynomials

$10(t-1)^2$, $t(t-1)^2$ and $(1+2t+t^2+3t^3)(t-1)^2$ annihilate the matrix A in Example 6.2. Is the characteristic polynomial one among the multiples of the minimal polynomial?

Theorem 6.4 *The minimal polynomial of a square matrix is unique; and it divides all annihilating polynomials of the matrix.*

Proof Let $A \in \mathbb{F}^{n \times n}$. Since the degree of a minimal polynomial is unique, let $p(t) = t^k + p_1(t)$ and $q(t) = t^k + q_1(t)$ be two minimal polynomials of A, where $p_1(t)$ and $q_1(t)$ are polynomials of degree less than k. Then $p(A) - q(A) = p_1(A) - q_1(A) = 0$.

The polynomial $r(t) = p_1(t) - q_1(t)$ has degree less than k and it annihilates A. Thus, $r(t)$ is the zero polynomial. In that case, $p(t) = q(t)$. This proves that the minimal polynomial of A is unique.

For the second statement, let $q(t)$ be the minimal polynomial of A. Let $p(t)$ be an annihilating polynomial of A. The degree of $p(t)$ is at least as large as that of $q(t)$. So, suppose the degree of $q(t)$ is k and the degree of $p(t)$ is $m \geq k$. By the division algorithm, we have $p(t) = s(t)q(t) + r(t)$, where $s(t)$ is a polynomial of degree $m - k$, and $r(t)$ is a polynomial of degree less than k. Then

$$0 = p(A) = s(A)\,q(A) + r(A) = 0 + r(A) = r(A).$$

That is, $r(t)$ is a polynomial of degree less than k that annihilates A. Hence, $r(t)$ is the zero polynomial. Therefore, $q(t)$ divides $p(t)$. ∎

Theorem 6.3 implies that

> the minimal polynomial of a matrix divides its characteristic polynomial.

This fact is sometimes mentioned as the Cayley–Hamilton theorem.

Exercises for Sect. 6.2

1. Compute A^{-1} and A^{10} where A is the matrix of Example 6.2.
2. Use the formula $(A - tI)\,\mathrm{adj}(A - tI) = \det(A - tI)I$ to give another proof of Cayley–Hamilton theorem. [Hint: For matrices $B_0, \ldots, B_{n-1}$, write $\mathrm{adj}(A - tI) = B_0 + B_1 t + \cdots + B_{n-1} t^{n-1}$.]
3. Let $A \in \mathbb{F}^{n \times n}$. Let $p(t)$ be a monic polynomial of degree n. If $p(A) = 0$, then does it follow that $p(t)$ is the characteristic polynomial of A?
4. Basing on Theorem 6.4, describe an algorithm for computing the minimal polynomial of a matrix.

6.3 Diagonalizability

Schur triangularization implies that each square matrix with complex entries is similar to an upper triangular matrix. Moreover, a square matrix with real entries is similar to an upper triangular real matrix provided all zeros of its characteristic polynomial are real. The upper triangular matrix similar to a given square matrix takes a better form when the matrix is hermitian.

Theorem 6.5 (Spectral theorem for hermitian matrices) *Each hermitian matrix is unitarily similar to a real diagonal matrix. And, each real symmetric matrix is orthogonally similar to a real diagonal matrix.*

Proof Let $A \in \mathbb{C}^{n \times n}$ be a hermitian matrix. Due to Schur triangularization, we have a unitary matrix P such that $D = P^*AP$ is upper triangular. Now,

$$D^* = P^*A^*P = P^*AP = D.$$

Since D is upper triangular and $D^* = D$, we see that D is a diagonal matrix. Comparing the diagonal entries in $D^* = D$, we see that all diagonal entries in D are real.

Further, if A is real symmetric, then all its eigenvalues are real. Due to Schur triangularization, the matrix P above can be chosen to be an orthogonal matrix so that $D = P^t AP$ is upper triangular. By a similar argument as above, it follows that D is a diagonal matrix with real entries. ∎

A matrix $A \in \mathbb{F}^{n \times n}$ is called **diagonalizable** if there exists an invertible matrix $P \in \mathbb{F}^{n \times n}$ such that $P^{-1}AP$ is a diagonal matrix. When $P^{-1}AP$ is a diagonal matrix, we say that A is **diagonalized by** P. In this language, the spectral theorem for hermitian matrices may be stated as follows:

Every hermitian matrix is unitarily diagonalizable; and every real symmetric matrix is orthogonally diagonalizable.

To see how the eigenvalues and eigenvectors are involved in the diagonalization process, let $A, P, D \in \mathbb{F}^{n \times n}$ be matrices, where

$$P = [v_1 \cdots v_n], \quad D = \mathrm{diag}(\lambda_1, \ldots, \lambda_n).$$

Then $Pe_j = v_j$ and $De_j = \lambda_j e_j$ for each $1 \leq j \leq n$. Suppose $AP = PD$. Then

$$Av_j = APe_j = PDe_j = P(\lambda_j e_j) = \lambda_j(Pe_j) = \lambda_j v_j \quad \text{for } 1 \leq j \leq n.$$

Conversely, suppose $Av_j = \lambda_j v_j$ for $1 \leq j \leq n$. Then

$$APe_j = Av_j = \lambda_j v_j = \lambda_j Pe_j = P(\lambda_j e_j) = PDe_j \quad \text{for } 1 \leq j \leq n.$$

That is, $AP = PD$. We summarize this as follows.

Observation 6.1 Let $A, P, D \in \mathbb{F}^{n \times n}$. For $j = 1, \ldots, n$, let v_j be the jth column of P. Let $D = \mathrm{diag}(\lambda_1, \ldots, \lambda_n)$. Then, $AP = PD$ iff $Av_j = \lambda_j v_j$ for each $j = 1, \ldots, n$.

For diagonalization of a matrix $A \in \mathbb{F}^{n \times n}$, we find the eigenvalues of A and the associated eigenvectors.

If $\mathbb{F} = \mathbb{R}$ and there exists an eigenvalue with nonzero imaginary part, then A cannot be diagonalized by a matrix with real entries.

Else, suppose we have n number of eigenvalues in $\mathbb{F}$ counting multiplicities.

If n linearly independent eigenvectors cannot be found, then A cannot be diagonalized.

Otherwise, we put the eigenvectors together as columns to form the matrix P; and $P^{-1}AP$ is a diagonalization of A. Further, by taking orthonormal eigenvectors, the matrix P can be made unitary or orthogonal.

Example 6.3 The matrix $A = \begin{bmatrix} 1 & -1 & -1 \\ -1 & 1 & -1 \\ -1 & -1 & 1 \end{bmatrix}$ is real symmetric. It has eigenvalues -1, 2 and 2, with associated orthonormal eigenvectors as

$$\begin{bmatrix} 1/\sqrt{3} \\ 1/\sqrt{3} \\ 1/\sqrt{3} \end{bmatrix}, \quad \begin{bmatrix} -1/\sqrt{2} \\ 1/\sqrt{2} \\ 0 \end{bmatrix}, \quad \begin{bmatrix} -1/\sqrt{6} \\ -1/\sqrt{6} \\ 2/\sqrt{6} \end{bmatrix}.$$

Thus, the diagonalizing orthogonal matrix is given by

$$P = \begin{bmatrix} 1/\sqrt{3} & -1/\sqrt{2} & -1/\sqrt{6} \\ 1/\sqrt{3} & 1/\sqrt{2} & -1/\sqrt{6} \\ 1/\sqrt{3} & 0 & 2/\sqrt{6} \end{bmatrix}.$$

We see that $P^{-1} = P^t$ and $P^{-1}AP = P^t AP = \begin{bmatrix} -1 & 0 & 0 \\ 0 & 2 & 0 \\ 0 & 0 & 2 \end{bmatrix}.$ $\square$

In fact, the spectral theorem holds for a bigger class of matrices. A matrix $A \in \mathbb{C}^{n \times n}$ is called a **normal matrix** iff $A^*A = AA^*$. All unitary matrices and all hermitian matrices are normal matrices. All diagonal matrices are normal matrices. In addition, a week converse to the last statement holds.

Theorem 6.6 *Each upper triangular normal matrix is diagonal.*

Proof Let $U \in \mathbb{C}^{n \times n}$ be an upper triangular matrix. If $n = 1$, then clearly U is a diagonal matrix. Suppose that each upper triangular normal matrix of order k is diagonal. Let U be an upper triangular normal matrix of order $k + 1$. Write U in a partitioned form as in the following:

$$U = \begin{bmatrix} R & u \\ 0 & a \end{bmatrix}$$

where $R \in \mathbb{C}^{k \times k}$, $u \in \mathbb{C}^{k \times 1}$, 0 is the zero row vector in $\mathbb{C}^{1 \times k}$, and $a \in \mathbb{C}$. Since U is normal,

$$U^*U = \begin{bmatrix} R^*R & R^*u \\ u^*R & u^*u + |a|^2 \end{bmatrix} = UU^* = \begin{bmatrix} RR^* + uu^* & \bar{a}\,u \\ a\,u^* & |a|^2 \end{bmatrix}.$$

It implies that $u^*u + |a|^2 = |a|^2$. That is, $u = 0$. Plugging $u = 0$ in the above equation, we see that $R^*R = RR^*$. Since R is upper triangular, by the induction hypothesis, R is a diagonal matrix. Then with $u = 0$, U is also a diagonal matrix. The proof is complete by induction. ∎

Using this result on upper triangular normal matrices, we can generalize the spectral theorem to normal matrices.

Theorem 6.7 (Spectral theorem for normal matrices) *A square matrix is unitarily diagonalizable iff it is a normal matrix.*

Proof Let $A \in \mathbb{C}^{n \times n}$. Let A be unitarily diagonalizable. Then there exist a unitary matrix P and a matrix $D = \mathrm{diag}(\lambda_1, \ldots, \lambda_n)$ such that $A = PDP^*$. Then $A^*A = PD^*DP^*$ and $AA^* = PDD^*P^*$. However,

$$D^*D = \mathrm{diag}\big(|\lambda_1|^2, \ldots, |\lambda_n|^2\big) = DD^*.$$

Therefore, $A^*A = AA^*$. So, A is a normal matrix.

Conversely let A be a normal matrix; so $A^*A = AA^*$. Due to Schur triangularization, let Q be a unitary matrix such that $Q^*AQ = U$, an upper triangular matrix. Since $Q^* = Q^{-1}$, the condition $A^*A = AA^*$ implies that $U^*U = UU^*$. By Theorem 6.6, U is a diagonal matrix. ∎

There can be non-normal matrices which are diagonalizable. For example, with

$$A = \begin{bmatrix} 1 & 0 & 0 \\ 4 & 3 & -2 \\ 2 & 1 & 0 \end{bmatrix}, \quad P = \begin{bmatrix} 0 & -1 & 0 \\ 1 & 5 & 2 \\ 1 & 3 & 1 \end{bmatrix}$$

we see that $A^*A \neq AA^*$, $P^*P \neq I$ but

$$P^{-1}AP = \begin{bmatrix} 1 & -1 & 2 \\ -1 & 0 & 0 \\ 2 & 1 & -1 \end{bmatrix} \begin{bmatrix} 1 & 0 & 0 \\ 4 & 3 & -2 \\ 2 & 1 & 0 \end{bmatrix} \begin{bmatrix} 0 & -1 & 0 \\ 1 & 5 & 2 \\ 1 & 3 & 1 \end{bmatrix} = \begin{bmatrix} 1 & 0 & 0 \\ 0 & 1 & 0 \\ 0 & 0 & 2 \end{bmatrix}.$$

Observe that in such a case, the diagonalizing matrix P is non-unitary.

In general, we have the following characterization of diagonalizability.

Theorem 6.8 *A matrix $A \in \mathbb{F}^{n \times n}$ is diagonalizable iff there exists a basis of $\mathbb{F}^{n \times 1}$ consisting of eigenvectors of A.*

Proof Let $A \in \mathbb{F}^{n \times n}$. Suppose A is diagonalizable. Then there exist an invertible matrix P and a diagonal matrix D such that $AP = PD$. Let v_j be the jth column of

P. By Observation 6.1, $Av_j = \lambda_j v_j$. As P is invertible, its columns $v_1, \ldots, v_n$ are nonzero; thus they are eigenvectors of A, and they form a basis for $\mathbb{F}^{n \times 1}$.

Conversely, suppose that $\{v_1, \ldots, v_n\}$ is a basis of $\mathbb{F}^{n \times 1}$ and that for each $j = 1, \ldots, n$, v_j is an eigenvector of A. Then there exists $\lambda_j \in \mathbb{F}$ such that $Av_j = \lambda_j v_j$. Construct $n \times n$ matrices

$$P = \begin{bmatrix} v_1 & \cdots & v_n \end{bmatrix}, \quad D = \mathrm{diag}(\lambda_1, \ldots, \lambda_n).$$

Observation 6.1 implies that $AP = PD$. Since the columns of P form a basis for $\mathbb{F}^{n \times 1}$, P is invertible. Therefore, A is diagonalizable. $\blacksquare$

In case, we have a basis $B = \{v_1, \ldots, v_n\}$ for $\mathbb{F}^{n \times 1}$ with $Av_j = \lambda_j v_j$, due to Theorems 3.9 and 3.10, the co-ordinate matrix $[A]_{B,B}$ is the diagonal matrix $\mathrm{diag}(\lambda_1, \ldots, \lambda_n)$. The question is, when are there n linearly independent eigenvectors of A? The spectral theorem for normal matrices provides a partial answer. Another partial answer on diagonalizability is as follows.

Theorem 6.9 *If an $n \times n$ matrix has n distinct eigenvalues, then it is diagonalizable.*

Proof Suppose $A \in \mathbb{C}^{n \times n}$ has n distinct eigenvalues $\lambda_1, \ldots, \lambda_n$ with corresponding eigenvectors $v_1, \ldots, v_n$. By Theorem 5.3, the vectors $v_1, \ldots, v_n$ are linearly independent, and thus form a basis for $\mathbb{C}^{n \times 1}$. Therefore, A is diagonalizable.

More directly, take $P = [v_1 \quad \cdots \quad v_n]$. Then P is invertible. By Observation 6.1, $P^{-1}AP = \mathrm{diag}(\lambda_1, \ldots, \lambda_n)$. $\blacksquare$

If λ is an eigenvalue of a matrix A with an associated eigenvector u, then $Au = \lambda u$; that is, $u \in N(A - \lambda I)$. The number of linearly independent solution vectors of $Au = \lambda u$ is $\dim(N(A - \lambda I))$, the nullity of $A - \lambda I$. This number has certain relation with diagonalizability of T.

Let λ be an eigenvalue of a matrix $A \in \mathbb{F}^{n \times n}$. The **geometric multiplicity** of λ is $\dim(N(A - \lambda I))$; and the **algebraic multiplicity** of λ is the largest natural number k such that $(t - \lambda)^k$ divides $\chi_A(t)$.

Observe that if λ is an eigenvalue of A, then its geometric multiplicity is the maximum number of linearly independent eigenvectors associated with λ; and its algebraic multiplicity is the number of times λ is a zero of the characteristic polynomial. Thus the algebraic multiplicity of an eigenvalue is often called its *multiplicity*.

Example 6.4 Let $A = \begin{bmatrix} 1 & 0 \\ 0 & 1 \end{bmatrix}$ and let $B = \begin{bmatrix} 1 & 1 \\ 0 & 1 \end{bmatrix}$.

The characteristic polynomials of both A and B are equal to $(t - 1)^2$. The eigenvalue $\lambda = 1$ has algebraic multiplicity 2 for both A and B.

For geometric multiplicities, we solve $Ax = x$ and $By = y$.

Now, $Ax = x$ gives $x = x$, which is satisfied by any vector in $\mathbb{F}^{2 \times 1}$. Thus, $N(A - I) = \mathbb{F}^{2 \times 1}$; consequently, the geometric multiplicity of the only eigenvalue 1 of A is $\dim(N(A - I)) = 2$.

For the matrix B, take $y = [a, b]^t$. Then $By = y$ gives $a + b = a$ and $b = b$. That is, $a = 0$; and b can be any scalar. For example, $[0, 1]^t$ is a solution. Then the geometric multiplicity of the eigenvalue 1 of B is $\dim(N(B - I)) = 1$. $\qquad\qquad$ □

Theorem 6.10 *The geometric multiplicity of an eigenvalue of a matrix is less than or equal to the algebraic multiplicity of that eigenvalue.*

Proof Let λ be an eigenvalue of a matrix A. Let ℓ be the geometric multiplicity and let k be the algebraic multiplicity of the eigenvalue λ. We have ℓ number of linearly independent eigenvectors of A associated with the eigenvalue λ, and no more. Extend the set of these eigenvectors to an ordered basis B of V. Let P be the matrix whose columns are the vectors in B. Due to Theorems 3.9 and 3.10, the matrix $M = P^{-1}AP = [A]_{B,B}$ may be written as

$$M = \begin{bmatrix} \lambda I_\ell & C \\ 0 & D \end{bmatrix}$$

for some matrices $C \in \mathbb{C}^{\ell \times (n-\ell)}$ and $D \in \mathbb{C}^{(n-\ell) \times (n-\ell)}$. Since A and M are similar, they have the same characteristic polynomial $X(t) = (\lambda - t)^\ell p(t)$ for some polynomial $p(t)$ of degree $n - \ell$.

But the zero λ of $X(t)$ is repeated k times. That is, $X(t) = (\lambda - t)^k q(t)$ for some polynomial $q(t)$ of which $(\lambda - t)$ is not a factor.

Notice that $\lambda - t$ may or may not be a factor of $p(t)$. In any case, $\ell \leq k$. ■

Theorem 6.11 *An $n \times n$ matrix A is diagonalizable iff the geometric multiplicity of each eigenvalue of A is equal to its algebraic multiplicity iff the sum of geometric multiplicities of all eigenvalues of A is n.*

Proof Let λ be an eigenvalue of an $n \times n$ matrix A with geometric multiplicity ℓ and algebraic multiplicity k. Let A be diagonalizable. Then we have a basis E of $\mathbb{F}^{n \times 1}$ which consists of eigenvectors of A, with respect to which the matrix of A is diagonal. In this diagonal matrix, there are exactly k number of entries equal to λ. In the basis E, there are k number of eigenvectors associated with λ. These eigenvectors corresponding to λ are linearly independent. There may be more number of linearly independent eigenvectors associated with λ, but no less. So, $\ell \geq k$. Then Theorem 6.10 implies that $\ell = k$.

Conversely, suppose that the geometric multiplicity of each eigenvalue is equal to its algebraic multiplicity. Then corresponding to each eigenvalue λ, we have exactly that many linearly independent eigenvectors as its algebraic multiplicity. Moreover, eigenvectors corresponding to distinct eigenvalues are linearly independent. Thus, collecting together the eigenvectors associated with all eigenvalues, we get n linearly independent eigenvectors which form a basis for $\mathbb{F}^{n \times 1}$. (See Problem 7.) Therefore, A is diagonalizable.

The second 'iff' statement follows since geometric multiplicity of each eigenvalue is at most its algebraic multiplicity. ■

In Example 6.4, we see that the geometric multiplicity of each (the only) eigen-value of the matrix A is equal to its algebraic multiplicity; both are 2. So, A is diagonalizable; in fact, it is already diagonal. But the geometric multiplicity of the (only) eigenvalue 1 of B is 1 while the algebraic multiplicity is 2. Therefore, B is not diagonalizable.

Exercises for Sect. 6.3

1. Diagonalize the given matrix, and then compute its fifth power:

(a) $\begin{bmatrix} 0 & 1 & 1 \\ 1 & 0 & 1 \\ 1 & 1 & 0 \end{bmatrix}$ (b) $\begin{bmatrix} 7 & -2 & 0 \\ -2 & 6 & -2 \\ 0 & -2 & 5 \end{bmatrix}$ (c) $\begin{bmatrix} 7 & -5 & 15 \\ 6 & -4 & 15 \\ 0 & 0 & 1 \end{bmatrix}$

2. Show that the following matrices are diagonalized by matrices in $\mathbb{R}^{3\times3}$.

(a) $\begin{bmatrix} 3/2 & -1/2 & 0 \\ -1/2 & 3/2 & 0 \\ 1/2 & -1/2 & 1 \end{bmatrix}$ (b) $\begin{bmatrix} 3 & -1/2 & -3/2 \\ 1 & 3/2 & 3/2 \\ -1 & 1/2 & 5/2 \end{bmatrix}$ (c) $\begin{bmatrix} 2 & -1 & 0 \\ -1 & 2 & 0 \\ 2 & 2 & 3 \end{bmatrix}$

3. If possible, diagonalize the following matrices:

(a) $\begin{bmatrix} 2 & 0 & 0 \\ 2 & 1 & 0 \\ 1 & 2 & -1 \end{bmatrix}$ (b) $\begin{bmatrix} 1 & -2 & 1 \\ 0 & 1 & 1 \\ 0 & 3 & -1 \end{bmatrix}$ (c) $\begin{bmatrix} 1 & 2 & 3 \\ 2 & 4 & 6 \\ -1 & -2 & -3 \end{bmatrix}$

4. Are the following matrices diagonalizable?

(a) $\begin{bmatrix} 2 & 3 \\ 6 & -1 \end{bmatrix}$ (b) $\begin{bmatrix} 1 & -10 & 0 \\ -1 & 3 & 1 \\ -1 & 0 & 4 \end{bmatrix}$ (c) $\begin{bmatrix} 2 & 1 & 0 & 0 \\ 0 & 2 & 0 & 0 \\ 0 & 0 & 2 & 0 \\ 0 & 0 & 0 & 5 \end{bmatrix}$

5. Check whether each of the following matrices is diagonalizable. If diagonalizable, find a basis of eigenvectors for $\mathbb{C}^{3\times1}$:

(a) $\begin{bmatrix} 1 & 1 & 1 \\ 1 & -1 & 1 \\ 1 & 1 & -1 \end{bmatrix}$ (b) $\begin{bmatrix} 1 & 1 & 1 \\ 0 & 1 & 1 \\ 0 & 0 & 1 \end{bmatrix}$ (c) $\begin{bmatrix} 1 & 0 & 1 \\ 1 & 1 & 0 \\ 0 & 1 & 1 \end{bmatrix}$

6. Find orthogonal or unitary diagonalizing matrices for the following:

(a) $\begin{bmatrix} 2 & 1 \\ 1 & 2 \end{bmatrix}$ (b) $\begin{bmatrix} 1 & 3+i \\ 3-i & 4 \end{bmatrix}$ (c) $\begin{bmatrix} -4 & -2 & 2 \\ -2 & -1 & 1 \\ 2 & 1 & -1 \end{bmatrix}$

7. Determine A^5 and a matrix B such that $B^2 = A$, where

(a) $A = \begin{bmatrix} 2 & 1 \\ -2 & -1 \end{bmatrix}$ (b) $A = \begin{bmatrix} 9 & -5 & 3 \\ 0 & 4 & 3 \\ 0 & 0 & 1 \end{bmatrix}$

8. If A is a normal matrix, then show that A^*, $I + A$ and A^2 are also normal.

6.4 Jordan Form

All matrices cannot be diagonalized since corresponding to an eigenvalue, there may not be sufficient number of linearly independent eigenvectors. Non-diagonalizability of a matrix $A \in \mathbb{F}^{n \times n}$ means that we cannot have a basis consisting of vectors v_j for $\mathbb{F}^{n \times 1}$ so that $Av_j = \lambda_j v_j$ for scalars λ_j. In that case, we would like to have a basis which would bring the matrix to a nearly diagonal form. Specifically, if possible, we would try to construct a basis $\{v_1, \dots, v_n\}$ such that

$$Av_j = \lambda_j v_j \quad \text{or} \quad Av_j = \lambda_j v_j + v_{j-1} \quad \text{for each } j.$$

Notice that the matrix similar to A with respect to such a basis would have λ_js on the diagonal, and possibly nonzero entries on the super diagonal (entries above the diagonal); all other entries being 0.

We will show that it is possible, by proving that there exists an invertible matrix P such that

$$P^{-1}AP = \text{diag}(J_1, \ J_2, \dots, J_k),$$

where each J_i is a block diagonal matrix of the form

$$J_i = \text{diag}(\tilde{J}_1(\lambda_i), \ \tilde{J}_2(\lambda_i), \dots, \tilde{J}_{s_i}(\lambda_i)),$$

for some s_i. Each matrix $\tilde{J}_j(\lambda_i)$ of order j here has the form

$$\tilde{J}_j(\lambda_i) = \begin{bmatrix} \lambda_i & 1 & & & \\ & \lambda_i & 1 & & \\ & & \ddots & \ddots & \\ & & & & 1 \\ & & & & \lambda_i \end{bmatrix}.$$

The missing entries are all 0. Such a matrix $\tilde{J}_j(\lambda_i)$ is called a **Jordan block** with diagonal entries λ_i. The order of the Jordan block is its order as a square matrix. Any matrix which is in the block diagonal form $\text{diag}(J_1, \ J_2, \dots, J_k)$ is said to be in **Jordan form**.

In writing Jordan blocks and Jordan forms, we do not show the zero entries for improving legibility. For instance, the following are possible Jordan form matrices of order 3 with all diagonal entries as 1:

$$\begin{bmatrix} 1 & & \\ & 1 & \\ & & 1 \end{bmatrix} \quad \begin{bmatrix} 1 & 1 & \\ & 1 & \\ & & 1 \end{bmatrix} \quad \begin{bmatrix} 1 & & \\ & 1 & 1 \\ & & 1 \end{bmatrix} \quad \begin{bmatrix} 1 & 1 & \\ & 1 & 1 \\ & & 1 \end{bmatrix}$$

Example 6.5 The following matrix is in Jordan form:

$$
\begin{bmatrix}
1 & & & & & & \\
 & 1 & & & & & \\
 & & 1 & 1 & & & \\
 & & & 1 & & & \\
 & & & & 2 & 1 & \\
 & & & & & 2 & 1 \\
 & & & & & & 2
\end{bmatrix}.
$$

It has three Jordan blocks for the eigenvalue 1 of which two are of order 1 and one of order 2; and it has one block of order 3 for the eigenvalue 2.

The eigenvalue 1 has geometric multiplicity 3, algebraic multiplicity 4, and the eigenvalue 2 has geometric multiplicity 1 and algebraic multiplicity 3.   □

In what follows, we will be using similarity transformations resulting out of elementary matrices. A similarity transformation that uses an elementary matrix $E[i, j]$ on a matrix A transforms A to $(E[i, j])^{-1} A E[i, j]$. Since $(E[i, j])^{-1} = (E[i, j])^t = E[i, j]$, the net effect of this transformation is described as follows:

$E[i, j]^{-1} A E[i, j] = E[i, j] A E[i, j]$ exchanges the ith and jth rows, and then exchanges the ith and the jth columns of A.

We will refer to this type of similarity transformations by the name **permutation similarity**.

Using the second type of elementary matrices, we have a similarity transformation $(E_\alpha[i])^{-1} A E_\alpha[i]$ for $\alpha \neq 0$. Since $(E_\alpha[i])^{-1} = E_{1/\alpha}[i]$ and $(E_\alpha[i])^t = E_\alpha[i]$, this similarity transformation has the following effect:

$(E_\alpha[i])^{-1} A E_\alpha[i] = E_{1/\alpha}[i] A E_\alpha[i]$ multiplies all entries in the ith row with $1/\alpha$, and then multiplies all entries in the ith column with α; thus keeping (i, i)th entry intact.

We will refer to this type of similarity transformation as **dilation similarity**. In particular, if A is such a matrix that its ith row has all entries 0 except the (i, i)th entry, and there is another entry on the ith column which is $\alpha \neq 0$, then $(E_\alpha[i])^{-1} A E_\alpha[i]$ is the matrix in which this α changes to 1 and all other entries are as in A.

The third type of similarity transformation applied on A yields the matrix $(E_\alpha[i, j])^{-1} A E_\alpha[i, j]$. Notice that $(E_\alpha[i, j])^{-1} = E_{-\alpha}[i, j]$ and $(E_\alpha[j, i])^t = E_\alpha[i, j]$. This similarity transformation changes a matrix A as described below:

$(E_\alpha[i, j])^{-1} A E_\alpha[i, j] = E_{-\alpha}[i, j] A E_\alpha[i, j]$ is obtained from A by subtracting α times the jth row from the ith row, and then adding α times the ith column to the jth column.

We name this type of similarity as a **combination similarity**.

In the formula for m_k, below we use the convention that for any matrix B of order n, B^0 is the identity matrix of order n.

Theorem 6.12 (Jordan form) *Each matrix $A \in \mathbb{C}^{n \times n}$ is similar to a matrix in Jordan form J, where the diagonal entries are the eigenvalues of A. For $1 \leq k \leq n$, if $m_k(\lambda)$ is the number of Jordan blocks of order k with diagonal entry λ, in J, then*

$$m_k(\lambda) = \operatorname{rank}((A - \lambda I)^{k-1}) - 2\operatorname{rank}((A - \lambda I)^k) + \operatorname{rank}((A - \lambda I))^{k+1}.$$

The Jordan form of A is unique up to a permutation of the blocks.

Proof First, we will show the existence of a Jordan form, and then we will come back to the formula m_k, which will show the uniqueness of a Jordan form up to a permutation of Jordan blocks.

Due to Schur triangularization, we assume that A is an upper triangular matrix, where the eigenvalues of A occur on the diagonal, and equal eigenvalues occur together. If $\lambda_1, \ldots, \lambda_k$ are the distinct eigenvalues of A, then our assumption means that A is an upper triangular matrix with diagonal entries, read from top left to bottom right, appear as

$$\lambda_1, \ldots, \lambda_1; \lambda_2, \ldots, \lambda_2; \ldots; \lambda_k, \ldots, \lambda_k.$$

Let n_i denote the number of times λ_i occurs in this list. First, we show that by way of a similarity transformation, A can be brought to the form

$$\operatorname{diag}(A_1, A_2, \ldots, A_k),$$

where each A_i is an upper triangular matrix of size $n_i \times n_i$ and each diagonal entry of A_i is λ_i. Our requirement is shown schematically as follows, where each such element marked x that is not inside the blocks A_i needs to be zeroed-out by a similarity transformation.

If such an x occurs as the (r, s)th entry in A, then $r < s$. Moreover, the corresponding diagonal entries a_{rr} and a_{ss} are eigenvalues of A occurring in different blocks A_i and A_j. Thus $a_{rr} \neq a_{ss}$. Further, all entries below the diagonals of A_i and of A_j are 0. We use a combination similarity to obtain

$$E_{-\alpha}[r, s] \, A \, E_\alpha[r, s] \quad \text{with} \quad \alpha = \frac{-x}{a_{rr} - a_{ss}}.$$

This similarity transformation subtracts α times the sth row from the rth row and then adds α times the rth column to the sth column. Since $r < s$, it changes the entries of A in the rth row to the right of the sth column, and the entries in the sth column above the rth row. Thus, the upper triangular nature of the matrix does not change. Further, it replaces the (r, s)th entry x with

$$a_{rs} + \alpha(a_{rr} - a_{ss}) = x + \frac{-x}{a_{rr} - a_{ss}}(a_{rr} - a_{ss}) = 0.$$

We use a sequence of such similarity transformations starting from the last row of A_{k-1} with smallest column index and ending in the first row with largest column index. Observe that an entry beyond the blocks, which was 0 previously can become nonzero after a single such similarity transformation. Such an entry will eventually be zeroed-out. Finally, each position which is not inside any of the k blocks $A_1, \ldots, A_k$ contains only 0. On completion of this stage, we end up with a matrix

$$\text{diag}(A_1, A_2, \ldots, A_k).$$

In the second stage, we focus on bringing each block A_i to the Jordan form. For notational convenience, write λ_i as a. If $n_i = 1$, then such an A_i is already in Jordan form. We use induction on the order n_i of A_i. Lay out the induction hypothesis that each such matrix of order $m - 1$ has a Jordan form. Suppose A_i has order m. Look at A_i in the following partitioned form:

$$A_i = \left[\begin{array}{c|c} B & u \\ \hline 0 & a \end{array}\right],$$

where B is the first $(m - 1) \times (m - 1)$ block, 0 is the zero row vector in $\mathbb{C}^{1 \times (m-1)}$, and u is a column vector in $\mathbb{C}^{(m-1) \times 1}$. By the induction hypothesis, there exists an invertible matrix Q such that $Q^{-1}BQ$ is in Jordan form; it looks like

$$Q^{-1}BQ = \begin{bmatrix} \boxed{B_1} & & & \\ & \boxed{B_2} & & \\ & & \ddots & \\ & & & \boxed{B_\ell} \end{bmatrix} \quad \text{where each } B_j = \begin{bmatrix} a & 1 & & \\ & a & 1 & \\ & & \ddots & \ddots \\ & & & & 1 \\ & & & & a \end{bmatrix}.$$

Then

$$\begin{bmatrix} Q & 0 \\ 0 & 1 \end{bmatrix}^{-1} A_i \begin{bmatrix} Q & 0 \\ 0 & 1 \end{bmatrix} = \begin{bmatrix} Q^{-1}BQ & Q^{-1}u \\ 0 & a \end{bmatrix} = \begin{bmatrix} a & * & & & b_1 \\ & a & * & & b_2 \\ & & \ddots & \ddots & \\ & & & * & b_{m-2} \\ & & & a & b_{m-1} \\ & & & & a \end{bmatrix}.$$

Call the above matrix as C. In the matrix C, the sequence of $*$'s on the super-diagonal, read from top left to right bottom, comprise a block of 1s followed by a 0, and then a block of 1s followed by a 0, and so on. The number of 1s depends on the sizes of B_1, B_2, etc. That is, when B_1 is over, and B_2 starts, we have a 0. Also, we have shown $Q^{-1}u$ as $[b_1 \cdots b_{m-1}]^t$. Our goal is to zero-out all b_js except b_{m-1} which may be made a 0 or 1.

In the next sub-stage, call it the third stage, we apply similarity transformations to zero-out (all or except one of) the entries $b_1, \ldots, b_{m-2}$. In any row of C, the entry above the diagonal (the $*$ there) is either 0 or 1. The $*$ is a 0 at the last row of each block B_j. We leave all such b's right now; they are to be tackled separately. So, suppose in the rth row, $b_r \neq 0$ and the $(r, r+1)$th entry (the $*$ above the diagonal entry) is a 1. We wish to zero-out each such b_r which is in the (r, m) position. For this purpose, we use a combination similarity to transform C to

$$E_{b_r}[r+1, m]\, C\, (E_{b_r}[r+1, m])^{-1} = E_{b_r}[r+1, m]\, C\, E_{-b_r}[r+1, m].$$

Observe that this matrix is obtained from C by adding b_r times the last row to the $(r+1)$th row, and then subtracting b_r times the $(r+1)$th column from the last column. Its net result is replacing the (r, m)th entry by 0, and keeping all other entries intact. Continuing this process of applying a suitable combination similarity transformation, each nonzero b_i with a corresponding 1 on the super-diagonal on the same row is reduced to 0. We then obtain a matrix, where all entries in the last column of C have been zeroed-out, without touching the entries at the last row of any of the blocks B_j. Write such entries as $c_1, \ldots, c_\ell$. Thus, at the end of third stage, A_i has been brought to the following from by similarity transformations:

$$
F := \begin{bmatrix}
\boxed{B_1} & & & & c_1 \\
& \boxed{B_2} & & & c_2 \\
& & \ddots & & \\
& & & \boxed{B_\ell} & c_\ell \\
& & & & a
\end{bmatrix}
$$

Notice that if B_j is a 1×1 block, then the corresponding entry c_j on the last column is already 0. In the next sub-stage, call it the fourth stage, we keep the nonzero c corresponding to the last block (the c entry with highest column index), and zero-out all other c's. Let B_q be the last block so that its corresponding c entry is $c_q \neq 0$ in the sth row. (It may not be c_ℓ; in that case, all of $c_{q+1}, \ldots, c_\ell$ are already 0.) We first make c_q a 1 by using a dilation similarity:

$$G := E_{1/c_q}[s]\, F\, E_{c_q}[s].$$

In G, the earlier c_q at (s, m)th position is now 1. Let B_p be any block other than B_q with $c_p \neq 0$ in the rth row. Our goal in this sub-stage, call it the fifth stage, is to zero-out c_p. We use two combination similarity transformations as shown below:

$$H := E_{-c_p}[r-1, s-1]\, E_{-c_p}[r, s]\, G\, E_{c_p}[r, s]\, E_{c_p}[r-1, s-1].$$

This similarity transformation brings c_p to 0 and keeps other entries intact. We do this for each such c_p. Thus in the mth column of H, we have only one nonzero entry

1 at (s, m)th position. If this happens to be at the last row, then we have obtained a
Jordan form. Otherwise, in this sub-stage (call it the seventh stage), we move this 1
to the $(s, s + 1)$th position by the following sequence of permutation similarities:

$$E[m - 1, m] \cdots E[s + 2, m]E[s + 1, m] \, H \, E[s + 1, m]E[s + 2, m] \cdots E[m - 1, m].$$

This transformation exchanges the rows and columns beyond the sth so that the 1 in
(s, m)th position moves to $(s, s + 1)$th position making up a block; and other entries
remain as they were earlier.

Here ends the proof by induction that each block A_i can be brought to a Jordan form
by similarity transformations. From a similarity transformation for A_i, a similarity
transformation can be constructed for the block diagonal matrix

$$\tilde{A} := \text{diag}(A_1, \ A_2, \ldots, A_k)$$

by putting identity matrices of suitable order and the similarity transformation for A_i
in a block form. As these transformations do not affect any other rows and columns
of $\tilde{A}$, a sequence of such transformations brings $\tilde{A}$ to its Jordan form, proving the
existence part in the theorem.

Toward the formula for m_k, let λ be an eigenvalue of A, and let $1 \le k \le n$.
Observe that $A - \lambda I$ is similar to $J - \lambda I$. Thus,

$$\text{rank}((A - \lambda I)^i) = \text{rank}((J - \lambda I)^i) \quad \text{for each } i.$$

Therefore, it is enough to prove the formula for J instead of A.

We use induction on n. In the basis case, $J = [\lambda]$. Here, $k = 1$ and $m_k = m_1 = 1$.
On the right hand side,

$$(J - \lambda I)^{k-1} = I = [1], \ (J - \lambda I)^k = [0]^1 = [0], \ (J - \lambda I)^{k+1} = [0]^2 = [0].$$

So, the formula holds for $n = 1$.

Lay out the induction hypothesis that for all matrices in Jordan form of order less
than n, the formula holds. Let J be a matrix of order n, which is in Jordan form. We
consider two cases.

Case 1: Let J have a single Jordan block corresponding to λ. That is,

$$J = \begin{bmatrix} \lambda & 1 & & & \\ & \lambda & 1 & & \\ & & \ddots & \ddots & \\ & & & & 1 \\ & & & & \lambda \end{bmatrix}, \quad J - \lambda I = \begin{bmatrix} 0 & 1 & & & \\ & 0 & 1 & & \\ & & \ddots & \ddots & \\ & & & & 1 \\ & & & & 0 \end{bmatrix}.$$

Here $m_1 = 0, \ m_2 = 0, \ldots, m_{n-1} = 0$ and $m_n = 1$. We see that $(J - \lambda I)^2$ has 1s on
the super-super-diagonal, and 0 elsewhere. Proceeding similarly for higher powers
of $J - \lambda I$, we see that their ranks are given by

$$\text{rank}(J - \lambda I) = n - 1, \text{rank}\big((J - \lambda I)^2\big) = n - 2, \ldots, \text{ rank}\big((J - \lambda I)^i\big) = n - i,$$
$$\text{rank}\big((J - \lambda I)^n\big) = 0, \text{ rank}\big((J - \lambda I)^{n+1}\big) = 0, \ldots$$

Then for $k < n$, $\text{rank}\big((J - \lambda I)^{k-1}\big) - 2\,\text{rank}\big((J - \lambda I)^k\big) + \text{rank}\big((J - \lambda I)^{k+1}\big)$
$$= (n - (k - 1)) - 2(n - k) + (n - k - 1) = 0.$$

And for $k = n$, $\text{rank}\big((J - \lambda I)^{k-1}\big) - 2\,\text{rank}\big((J - \lambda I)^k\big) + \text{rank}\big((J - \lambda I)^{k+1}\big)$
$$= (n - (n - 1)) - 2 \times 0 + 0 = 1 = m_n.$$

Case 2: Suppose J has more than one Jordan block corresponding to λ. The first Jordan block in J corresponds to λ and has order r for some $r < n$. Then $J - \lambda I$ can be written in block form as

$$J - \lambda I = \begin{bmatrix} C & 0 \\ 0 & D \end{bmatrix},$$

where C is the Jordan block of order r with diagonal entries as 0, and D is the matrix of order $n - r$ in Jordan form consisting of other blocks of $J - \lambda I$. Then, for any j,

$$(J - \lambda I)^j = \begin{bmatrix} C^j & 0 \\ 0 & D^j \end{bmatrix}.$$

Therefore, $\text{rank}(J - \lambda I)^j = \text{rank}(C^j) + \text{rank}(D^j)$. Write $m_k(C)$ and $m_k(D)$ for the number of Jordan blocks of order k for the eigenvalue λ that appear in C, and in D, respectively. Then

$$m_k = m_k(C) + m_k(D).$$

By the induction hypothesis,

$$m_k(C) = \text{rank}(C^{k-1}) - 2\,\text{rank}(C^k) + \text{rank}(C)^{k+1},$$
$$m_k(D) = \text{rank}(D^{k-1}) - 2\,\text{rank}(D^k) + \text{rank}(D)^{k+1}.$$

It then follows that

$$m_k = \text{rank}((J - \lambda I)^{k-1}) - 2\,\text{rank}((J - \lambda I)^k) + \text{rank}((J - \lambda I))^{k+1}.$$

Since the number of Jordan blocks of order k corresponding to each eigenvalue of A is uniquely determined, the Jordan form of A is also uniquely determined up to a permutation of blocks.   ∎

To obtain a Jordan form of a given matrix, we may use the construction of similarity transformations as used in the proof of Theorem 6.12, or we may use the formula for m_k as given there. We illustrate these methods in the following examples.

Example 6.6 Let $A =$

$$\begin{bmatrix} 2 & 1 & 0 & 0 & 0 & 1 & 0 & ② & 0 \\ & 2 & 0 & 0 & 0 & 3 & 0 & 0 & ① \\ & & 2 & 1 & 0 & 0 & ② & 0 & 0 \\ & & & 2 & 0 & 2 & 0 & 0 & 0 \\ & & & & 2 & 0 & 0 & 0 & 0 \\ & & & & & 2 & 0 & 0 & 0 \\ & & & & & & 3 & 1 & 1 \\ & & & & & & & 3 & 1 \\ & & & & & & & & 3 \end{bmatrix}.$$

This is an upper triangular matrix. Following the proof of Theorem 6.12, we first zero-out the circled entries, starting from the entry on the third row. Here, the row index is $r = 3$, the column index is $s = 7$, the eigenvalues are $a_{rr} = 2$, $a_{ss} = 3$, and the entry to be zeroed-out is $x = 2$. Thus, $\alpha = -2/(2-3) = 2$. We use an appropriate combination similarity to obtain

$$M_1 = E_{-2}[3, 7] \, A \, E_2[3, 7].$$

That is, in A, we replace $row(3)$ with $row(3) - 2 \times row(7)$ and then replace $col(7)$ with $col(7) + 2 \times col(3)$. It leads to

$$M_1 = \begin{bmatrix} 2 & 1 & 0 & 0 & 0 & 1 & 0 & ② & 0 \\ & 2 & 0 & 0 & 0 & 3 & 0 & 0 & ① \\ & & 2 & 1 & 0 & 0 & 0 & -2 & 0 \\ & & & 2 & 0 & 2 & 0 & 0 & 0 \\ & & & & 2 & 0 & 0 & 0 & 0 \\ & & & & & 2 & 0 & 0 & 0 \\ & & & & & & 3 & 1 & 1 \\ & & & & & & & 3 & 1 \\ & & & & & & & & 3 \end{bmatrix}.$$

Notice that the similarity transformation brought in a new nonzero entry such as -2 in $(3, 8)$ position. But its column index has increased. Looking at the matrix afresh, we must zero-out this entry first. The suitable combination similarity yields

$$M_2 = E_2[3, 8] \, M_1 \, E_{-2}[3, 8]$$

which replaces $row(3)$ with $row(3) + 2 \times row(8)$ and then replaces $col(8)$ with $col(8) - 2 \times col(3)$. Verify that it zeroes-out the entry -2 but introduces 2 at $(3, 9)$ position. Once more, we use a combination similarity to obtain

$$M_3 = E_{-2}[3, 9] \, M_2 \, E_2[3, 9]$$

replacing $row(3)$ with $row(3) - 2 \times row(9)$ and then replacing $col(9)$ with $col(9) + 2 \times col(3)$. Now,

$$M_3 = \begin{bmatrix} 2 & 1 & 0 & 0 & 0 & 1 & 0 & ② & 0 \\ & 2 & 0 & 0 & 0 & 3 & 0 & 0 & ① \\ & & 2 & 1 & 0 & 0 & 0 & 0 & 0 \\ & & & 2 & 0 & 2 & 0 & 0 & 0 \\ & & & & 2 & 0 & 0 & 0 & 0 \\ & & & & & 2 & 0 & 0 & 0 \\ & & & & & & 3 & 1 & 1 \\ & & & & & & & 3 & 1 \\ & & & & & & & & 3 \end{bmatrix}.$$

Similar to the above, we use the combination similarities to reduce M_3 to M_4, where

$$M_4 = E_{-1}[2, 9]\, M_3\, E_1[2, 9].$$

To zero-out the encircled 2, we use a suitable combination similarity, and get

$$M_5 = E_{-2}[1, 8]\, M_4\, E_2[1, 8].$$

It zeroes-out the encircled 2 but introduces -2 at $(1, 9)$ position. Once more, we use a suitable combination similarity to obtain

$$M_6 = E_2[1, 9]\, M_5\, E_{-2}[1, 9] = \begin{bmatrix} 2 & 1 & 0 & 0 & 0 & ① \\ & 2 & 0 & 0 & 0 & 3 \\ & & 2 & 1 & 0 & 0 \\ & & & 2 & 0 & 2 \\ & & & & 2 & 0 \\ & & & & & 2 \\ & & & & & & 3 & 1 & 1 \\ & & & & & & & 3 & 1 \\ & & & & & & & & 3 \end{bmatrix}.$$

Now, the matrix M_6 is in block diagonal form. We focus on each of the blocks, though we will be working with the whole matrix. We consider the block corresponding to the eigenvalue 2 first. Since this step is inductive we scan this block from the top left corner. The 2×2 principal sub-matrix of this block is already in Jordan form. The 3×3 principal sub-matrix is also in Jordan form. We see that the principal sub-matrix of size 4×4 and 5×5 is also in Jordan form, but the 6×6 sub-matrix, which is the block itself is not in Jordan form.

We wish to bring the sixth column to its proper shape. Recall that our strategy is to zero out all those entries on the sixth column which are opposite to a 1 on the super-diagonal of this block. There is only one such entry, which is encircled in M_6 above.

The row index of this entry is $r = 1$, its column index is $m = 6$, and the entry itself is $b_r = 1$. We use a combination similarity to obtain

$$M_7 = E_1[2,6]\, M_6\, E_{-1}[2,6] = \begin{bmatrix} 2 & 1 & 0 & 0 & 0 & 0 \\ & 2 & 0 & 0 & 0 & 5 \\ & & 2 & 1 & 0 & 0 \\ & & & 2 & 0 & 2 \\ & & & & 2 & 0 \\ & & & & & 2 \\ & & & & & & 3 & 1 & 1 \\ & & & & & & & 3 & 1 \\ & & & & & & & & 3 \end{bmatrix}.$$

Next, among the nonzero entries 5 and 2 at the positions $(2,6)$ and $(4,6)$, we wish to zero-out the 5 and keep 2 as the row index of 2 is higher. First, we use a dilation similarity to make this entry 1 as in the following:

$$M_8 = E_{1/2}[4]\, M_7\, E_2[4].$$

It replaces $row(4)$ with $1/2$ times itself and then replaces $col(4)$ with 2 times itself, thus making $(4,6)$th entry 1 and keeping all other entries intact. Next, we zero-out the 5 on $(2,4)$ position by using the two combination similarities. Here, $c_p = 5$, $r = 2$, $s = 4$; thus

$$M_9 = E_{-5}[1,3]\, E_{-5}[2,4]\, M_8\, E_5[2,4]\, E_5[1,3] = \begin{bmatrix} 2 & 1 & 0 & 0 & 0 & 0 \\ & 2 & 0 & 0 & 0 & 0 \\ & & 2 & 1 & 0 & 0 \\ & & & 2 & 0 & ① \\ & & & & 2 & 0 \\ & & & & & 2 \\ & & & & & & 3 & 1 & 1 \\ & & & & & & & 3 & 1 \\ & & & & & & & & 3 \end{bmatrix}.$$

Here, M_9 has been obtained from M_8 by replacing $row(2)$ with $row(2) - 5 \times row(4)$, $col(4)$ with $col(4) + 5 \times col(2)$, $row(1)$ with $row(1) - 5 \times row(3)$, and then $col(3)$ with $col(3) + 5 \times col(1)$.

Next, we move this encircled 1 to $(4,5)$ position by similarity. Here, $s = 4$, $m = 6$. Thus, the sequence of permutation similarities boils down to only one, i.e., exchanging $row(5)$ with $row(6)$ and then exchanging $col(6)$ with $col(5)$. Observe that we would have to use more number of permutation similarities if the difference between m and s is more than 2. It gives

$$M_{10} = E[5, 6] \, M_9 \, E[5, 6] = \begin{bmatrix} 2 & 1 & 0 & 0 & 0 & 0 & & & \\ & 2 & 0 & 0 & 0 & 0 & & & \\ & & 2 & 1 & 0 & 0 & & & \\ & & & 2 & 1 & 0 & & & \\ & & & & 2 & 0 & & & \\ & & & & & 2 & & & \\ & & & & & & 3 & 1 & 1 \\ & & & & & & & 3 & 1 \\ & & & & & & & & 3 \end{bmatrix}.$$

Now, the diagonal block corresponding to the eigenvalue 2 is in Jordan form. We focus on the other block corresponding to 3. Here, $(7, 9)$th entry which contains a 1 is to be zeroed-out. This entry is opposite to a 1 on the super-diagonal. We use a combination similarity. Here, the row index is $r = 7$, the column index $m = 9$, and the entry is $b_r = 1$. Thus, we have Jordan form as

$$M_{11} = E_1[8, 9] \, M_{10} \, E_{-1}[8, 9] = \begin{bmatrix} 2 & 1 & & & & & & & \\ & 2 & & & & & & & \\ & & 2 & 1 & & & & & \\ & & & 2 & 1 & & & & \\ & & & & 2 & & & & \\ & & & & & 2 & & & \\ & & & & & & 3 & 1 & \\ & & & & & & & 3 & 1 \\ & & & & & & & & 3 \end{bmatrix}. \qquad \square$$

Example 6.7 Consider the matrix A of Example 6.6. Here, we compute the number m_k of Jordan blocks of size k corresponding to each eigenvalue. For this purpose, we require the ranks of the matrices $(A - \lambda I)^k$ for successive k, and for each eigenvalue λ of A. We see that A has two eigenvalues 2 and 3.

For the eigenvalue 2,

$$\text{rank}(A - 2I)^0 = \text{rank}(I) = 9, \ \text{rank}(A - 2I) = 6, \ \text{rank}(A - 2I)^2 = 4,$$
$$\text{rank}(A - 2I)^{3+k} = 3 \quad \text{for } k = 0, \ 1, \ 2, \ldots$$

For the eigenvalue 3,

$$\text{rank}(A - 3I)^0 = \text{rank}(I) = 9, \ \text{rank}(A - 3I) = 8, \ \text{rank}(A - 3I)^2 = 7,$$
$$\text{rank}(A - 3I)^{3+k} = 6 \quad \text{for } k = 0, \ 1, \ 2, \ldots$$

Using the formula for $m_k(\lambda)$, we obtain

$$m_1(2) = 9 - 2 \times 6 + 4 = 1, \quad m_2(2) = 6 - 2 \times 4 + 3 = 1,$$
$$m_3(2) = 4 - 2 \times 3 + 3 = 1, \quad m_{3+k}(2) = 3 - 2 \times 3 + 3 = 0.$$

$$m_1(3) = 9 - 2 \times 8 + 7 = 0, \quad m_2(3) = 8 - 2 \times 7 + 6 = 0,$$
$$m_3(3) = 7 - 2 \times 6 + 6 = 1, \quad m_{3+k}(3) = 6 - 2 \times 6 + 6 = 0.$$

Therefore, in the Jordan form of A, there is one Jordan block of size 1, one of size 2 and one of size 3 with eigenvalue 2, and one block of size 3 with eigenvalue 3. From this information, we see that the Jordan form of A is uniquely determined up to any rearrangement of the blocks. Check that M_{11} as obtained in Example 6.6 is one such Jordan form of A.   □

Suppose that a matrix $A \in \mathbb{C}^{n \times n}$ has a Jordan form $J = P^{-1}AP$, in which the first Jordan block is of size k with diagonal entries as λ. If $P = [v_1 \ \cdots \ v_n]$, then $AP = PJ$ implies that

$$A(v_1) = \lambda v_1, \ A(v_2) = v_1 + \lambda v_2, \ldots, A(v_k) = v_{k-1} + \lambda v_k.$$

If the next Jordan block in J has diagonal entries as μ (which may or may not be equal to λ), then we have $Av_{k+1} = \mu v_{k+1}$, $Av_{k+2} = v_{k+1} + \mu v_{k+2}, \ldots$, and so on.

The list of vectors $v_1, \ldots, v_k$ above is called a *Jordan string that starts with v_1 and ends with v_k*. The number k is called the *length* of the Jordan string. In such a Jordan string, we see that

$$v_1 \in N(A - \lambda I), \ v_2 \in N(A - \lambda I)^2, \ldots, v_k \in N(A - \lambda I)^k.$$

Any vector in $N((A - \lambda I)^j)$, for some j, is called a **generalized eigenvector** corresponding to the eigenvalue λ of A.

The columns of P are all generalized eigenvectors of A corresponding to the eigenvalues of A. These generalized eigenvectors form a basis for $\mathbb{F}^{n \times 1}$. Such a basis is called a *Jordan basis*. The coordinate matrix of A with respect to the Jordan basis is the Jordan form J of A.

The Jordan basis consists of Jordan strings. Each Jordan string starts with an eigenvector of A, such as v_1 above. If λ is an eigenvalue of A having geometric multiplicity γ, then there are exactly γ number of Jordan strings in the Jordan basis corresponding to the eigenvalue λ. Thus, there are exactly γ number of Jordan blocks in J with diagonal entries as λ. The size of any such block is equal to the length of the corresponding Jordan string.

The uniqueness of a Jordan form can be made exact by first ordering the eigenvalues of A and then arranging the blocks corresponding to each eigenvalue (which now appear together on the diagonal) in some order, say in ascending order of their size. In doing so, the Jordan form of any matrix becomes unique. Such a Jordan form is called the **Jordan canonical form** of a matrix. It then follows that if two matrices are similar, then they have the same Jordan canonical form. Moreover, uniqueness also implies that two dissimilar matrices will have different Jordan canonical forms. Therefore, Jordan form characterizes similarity of matrices.

As an application of Jordan form, we will show that each matrix is similar to its transpose. Suppose $J = P^{-1}AP$. Now, $J^t = P^t A^t (P^{-1})^t = P^t A^t (P^t)^{-1}$. That is, A^t is similar to J^t. Thus, it is enough to show that J^t is similar to J. First, let us see it for a single Jordan block. For a Jordan block J_λ, consider the matrix Q of the same order as in the following:

$$
J_\lambda = \begin{bmatrix} \lambda & 1 & & & \\ & \lambda & 1 & & \\ & & \ddots & \ddots & \\ & & & & 1 \\ & & & & \lambda \end{bmatrix}, \quad Q = \begin{bmatrix} & & & & 1 \\ & & & 1 & \\ & & \cdot^{\cdot^{\cdot}} & & \\ & 1 & & & \\ 1 & & & & \end{bmatrix}.
$$

In the matrix Q, the entries on the *anti-diagonal* are all 1 and all other entries are 0. We see that $Q^2 = I$. Thus, $Q^{-1} = Q$. Further,

$$
Q^{-1} J_\lambda \, Q = Q \, J_\lambda \, Q = (J_\lambda)^t.
$$

Therefore, each Jordan block is similar to its transpose. Now, construct a matrix R by putting matrices such as Q as its blocks matching the orders of each Jordan block in J. Then it follows that $R^{-1} J R = J^t$.

Jordan form guarantees that one can always choose m linearly independent generalized eigenvectors corresponding to the eigenvalue λ, where m is the algebraic multiplicity of λ. Moreover, the following is guaranteed:

If the linear system $(A - \lambda I)^k x = 0$ has $r < m$ number of linearly independent solutions, then $(A - \lambda I)^{k+1} = 0$ has at least $r + 1$ number of linearly independent solutions.

This result is more useful in computing the exponential of a square matrix rather than using the Jordan form explicitly. See Sect. 7.6 for details.

Exercises for Sect. 6.4

1. Determine the Jordan forms of the following matrices:

   (a) $\begin{bmatrix} 0 & 0 & 0 \\ 1 & 0 & 0 \\ 2 & 1 & 0 \end{bmatrix}$ (b) $\begin{bmatrix} 0 & 0 & 1 \\ 0 & 1 & 0 \\ 1 & 0 & 0 \end{bmatrix}$ (c) $\begin{bmatrix} -2 & -1 & -3 \\ 4 & 3 & 3 \\ -2 & 1 & -1 \end{bmatrix}$

2. Determine the matrix $P \in \mathbb{C}^{3 \times 3}$ such that $P^{-1}AP$ is in Jordan form, where A is the matrix in Exercise 1(c).

3. Let A be a 7×7 matrix with characteristic polynomial $(t - 2)^4 (3 - t)^3$. It is known that in the Jordan form of A, the largest blocks for both the eigenvalues are of order 2. Show that there are only two possible Jordan forms for A; and determine those Jordan forms.

4. Let A be a 5×5 matrix whose first and second rows are, respectively, $[0, 1, 1, 0, 1]$ and $[0, 0, 1, 1, 1]$; and all other rows are zero rows. What is the Jordan form of A?

5. Let A be an $n \times n$ a lower triangular matrix with each diagonal entry as 1, each sub-diagonal entry (just below each diagonal entry) as 1, and all other entries being 0. What is the Jordan form of A?

6. Let A be an $n \times n$ matrix, where the diagonal entries are $1, 2, \ldots, n$, from top left to right bottom, the super-diagonal entries are all 1, and all other entries are 0. What is the Jordan form of A?

7. What is the Jordan form of the $n \times n$ matrix whose each row is equal to $[1, 2, \cdots, n]$?

8. Let λ be an eigenvalue of the $n \times n$ matrix A. Suppose for each $k \in \mathbb{N}$, we know the number m_k. Show that for each j, both $\operatorname{rank}(A - \lambda I)^j$ and $\operatorname{null}(A - \lambda I)^j$ are uniquely determined.

9. Prove that two matrices $A, B \in \mathbb{C}^{n \times n}$ are similar iff they have the same eigenvalues, and for each eigenvalue λ, $\operatorname{rank}(A - \lambda I)^k = \operatorname{rank}(B - \lambda I)^k$ for each $k \in \mathbb{N}$.

10. Show that two matrices $A, B \in \mathbb{C}^{n \times n}$ are similar iff they have the same eigenvalues, and for each eigenvalue λ, $\operatorname{null}(A - \lambda I)^k = \operatorname{null}(B - \lambda I)^k$ for each $k \in \mathbb{N}$.

6.5 Singular Value Decomposition

Given an $m \times n$ matrix A with complex entries, there are two hermitian matrices that can be constructed naturally from it, namely A^*A and AA^*. We wish to study the eigenvalues and eigenvectors of these matrices and their relations to certain parameters associated with A. We will see that these concerns yield a factorization of A.

All eigenvalues of the hermitian matrix $A^*A \in \mathbb{C}^{n \times n}$ are real. If $\lambda \in \mathbb{R}$ is such an eigenvalue with an associated eigenvector $v \in \mathbb{C}^{n \times 1}$, then $A^*Av = \lambda v$ implies that

$$\lambda \|v\|^2 = \lambda v^*v = v^*(\lambda v) = v^*A^*Av = (Av)^*(Av) = \|Av\|^2.$$

Since $\|v\| > 0$, we see that $\lambda \geq 0$. The eigenvalues of A^*A can thus be arranged in a decreasing list

$$\lambda_1 \geq \lambda_2 \geq \cdots \geq \lambda_r > 0 = \lambda_{r+1} = \cdots = \lambda_n$$

for some r with $0 \leq r \leq n$. Notice that $\lambda_1, \ldots, \lambda_r$ are all positive and the rest are all equal to 0. In the following we relate this r with $\operatorname{rank}(A)$. Of course, we could have considered AA^* instead of A^*A.

Let $A \in \mathbb{C}^{m \times n}$. Let $\lambda_1 \geq \cdots \geq \lambda_n \geq 0$ be the n eigenvalues of A^*A. The non-negative square roots of these real numbers are called the **singular values** of A. Conventionally, we denote the singular values of A by s_i. The eigenvalues of A^*A are then denoted by $s_1^2 \geq \cdots \geq s_n^2 \geq 0$.

Theorem 6.13 *Let* $A \in \mathbb{C}^{m \times n}$. *Then* $\operatorname{rank}(A) = \operatorname{rank}(A^*A) = \operatorname{rank}(AA^*) = \operatorname{rank}(A^*) = $ *the number of positive singular values of* A.

Proof Let $v \in \mathbb{C}^{n \times 1}$. If $Av = 0$, then $A^*Av = 0$. Conversely, if $A^*Av = 0$, then $v^*A^*Av = 0$. It implies that $\|Av\|^2 = 0$, giving $Av = 0$. Therefore, $N(A^*A) = N(A)$. It follows that $\text{null}(A^*A) = \text{null}(A)$. By the rank nullity theorem, we conclude that $\text{rank}(A^*A) = \text{rank}(A)$.

Consider A^* instead of A to obtain $\text{rank}(AA^*) = \text{rank}((A^*)^*A^*) = \text{rank}(A^*)$. Since $\text{rank}(A^t) = \text{rank}(A) = \text{rank}(\overline{A})$, $\text{rank}(A^*) = \text{rank}(A)$.

Next, let $s_1 \geq s_2 \geq \cdots \geq s_r > 0 = s_{r+1} = \cdots = s_n$ be the singular values of A. That is, there are exactly r number of positive singular values of A. By the spectral theorem, the hermitian matrix A^*A is unitarily diagonalizable. So, there exists a unitary matrix Q such that

$$Q^*(A^*A)Q = \text{diag}(s_1^2, \ldots, s_r^2, 0, \ldots, 0).$$

Therefore, $\text{rank}(A^*A)$ is equal to the rank of the above diagonal matrix; and that is equal to r. This completes the proof. ∎

We remark that the number of nonzero eigenvalues of a matrix, counting multiplicities, need not be same as the rank of the matrix. For instance, the matrix

$$\begin{bmatrix} 0 & 1 \\ 0 & 0 \end{bmatrix}$$

has no nonzero eigenvalues, but its rank is 1.

Suppose $\lambda > 0$ is an eigenvalue of A^*A with an associated eigenvector v. Then $A^*Av = \lambda v$ implies $(AA^*)(Av) = \lambda(Av)$. Since $\lambda v \neq 0$, $Av \neq 0$. Thus, λ is also an eigenvalue of AA^* with an associated eigenvector Av.

Similarly, if $\lambda > 0$ is an eigenvalue of AA^*, then it follows that the same λ is an eigenvalue of A^*A. That is,

a positive real number is an eigenvalue of A^*A iff it is an eigenvalue of AA^*.

From Theorem 6.13, it follows that A and A^* have the same r number of positive singular values, where $r = \text{rank}(A) = \text{rank}(A^*)$. Further, A has $n - r$ number of zero singular values, whereas A^* has $m - r$ number of zero singular values. In addition, if $A \in \mathbb{C}^{n \times n}$ is hermitian and has eigenvalues $\lambda_1, \ldots, \lambda_n$, then its singular values are $|\lambda_1|, \ldots, |\lambda_n|$.

Analogous to the factorization of A^*A, we have one for A itself.

Theorem 6.14 (SVD) *Let $A \in \mathbb{C}^{m \times n}$ be of rank r. Let $s_1 \geq \ldots \geq s_r$ be the positive singular values of A. Write $S = \text{diag}(s_1, \ldots, s_r) \in \mathbb{C}^{r \times r}$ and $\Sigma = \begin{bmatrix} S & 0 \\ 0 & 0 \end{bmatrix} \in \mathbb{C}^{m \times n}$. Then there exist unitary matrices $P = [u_1 \quad \cdots \quad u_m] \in \mathbb{C}^{m \times m}$ and $Q = [v_1 \quad \cdots \quad v_n] \in \mathbb{C}^{n \times n}$ such that the following are true:*

(1) $A = P \Sigma Q^* = s_1 u_1 v_1^* + \cdots + s_r u_r v_r^*$.
(2) (a) For $1 \leq i \leq r$, $AA^*u_i = s_i^2 AA^*u_i$; and for $r < i \leq m$, $AA^*u_i = 0$.
(b) For $1 \leq j \leq r$, $A^*Av_j = s_j^2 v_j$; and for $r < j \leq n$, $A^*Av_j = 0$.

(3) For $1 \le i \le r$, $u_i = (s_i)^{-1}Av_i$ and $v_i = (s_i)^{-1}A^*u_i$.
(4) (a) $\{u_1, \ldots, u_r\}$ is an orthonormal basis of $R(A)$.
(b) $\{v_1, \ldots, v_r\}$ is an orthonormal basis of $R(A^*)$.
(c) $\{v_{r+1}, \ldots, v_n\}$ is an orthonormal basis of $N(A)$.
(d) $\{u_{r+1}, \ldots, u_m\}$ is an orthonormal basis of $N(A^*)$.

Proof All positive singular values of A are $s_1 \ge \cdots \ge s_r$. Thus, the eigenvalues of A^*A are $s_1^2 \ge \cdots \ge s_r^2$, and 0 repeated $n - r$ times. Since $A^*A \in \mathbb{C}^{n \times n}$ is hermitian, there exists a unitary matrix $Q \in \mathbb{C}^{n \times n}$ such that

$$Q^*A^*AQ = \text{diag}(s_1^2, \ldots, s_r^2, 0, \ldots, 0) = \begin{bmatrix} S^2 & 0 \\ 0 & 0 \end{bmatrix},$$

where the columns of Q are eigenvectors of A^*A. Define

$$D = \text{diag}(s_1^{-1}, \ldots, s_r^{-1}, 1, \ldots, 1) = \begin{bmatrix} S^{-1} & 0 \\ 0 & I_{n-r} \end{bmatrix}.$$

Then $D^* = D$, and

$$(AQD)^*(AQD) = D^*(Q^*A^*AQ)D = \begin{bmatrix} S^{-1} & 0 \\ 0 & I_{n-r} \end{bmatrix}\begin{bmatrix} S^2 & 0 \\ 0 & 0 \end{bmatrix}\begin{bmatrix} S^{-1} & 0 \\ 0 & I_{n-r} \end{bmatrix} = \begin{bmatrix} I_r & 0 \\ 0 & 0 \end{bmatrix}.$$

Therefore, the first r columns, say, $u_1, \ldots, u_r$, of the $m \times n$ matrix AQD form an orthonormal set in $\mathbb{C}^{m \times 1}$; and the other columns of AQD are zero columns. Extend this orthonormal set $\{u_1, \ldots, u_r\}$ to an orthonormal basis $\{u_1, \ldots, u_r, u_{r+1}, \ldots, u_m\}$ for $\mathbb{C}^{m \times 1}$. Construct the matrices

$$P_1 = \begin{bmatrix} u_1 & \cdots & u_r \end{bmatrix}, \quad P_2 = \begin{bmatrix} u_{r+1} & \cdots & u_m \end{bmatrix}, \quad P = \begin{bmatrix} P_1 & P_2 \end{bmatrix}.$$

(1) We find that $P \in \mathbb{C}^{m \times m}$ is unitary, and $AQD = \begin{bmatrix} P_1 & 0 \end{bmatrix}$. We already have the unitary matrix $Q \in \mathbb{C}^{n \times n}$. Hence

$$A = (AQD)D^{-1}Q^* = \begin{bmatrix} P_1 & 0 \end{bmatrix}\begin{bmatrix} S & 0 \\ 0 & I_{n-r} \end{bmatrix}Q^* = \begin{bmatrix} P_1 & P_2 \end{bmatrix}\begin{bmatrix} S & 0 \\ 0 & 0 \end{bmatrix}Q^* = P\Sigma Q^*.$$

Also, we see that

$$P\Sigma Q = \begin{bmatrix} u_1 & \cdots & u_m \end{bmatrix}\begin{bmatrix} s_1 & & & & \\ & s_2 & & & \\ & & \ddots & & \\ & & & s_r & \\ & & & & 0 \end{bmatrix}\begin{bmatrix} v_1^* \\ \vdots \\ v_n^* \end{bmatrix} = s_1u_1v_1^* + \cdots + s_ru_rv_r^*.$$

(2) (a) $AA^*P = P \Sigma Q^* Q \Sigma^* P^* P = \Sigma \Sigma^* P$. The matrix $\Sigma \Sigma^*$ is a diagonal matrix with diagonal entries as $s_1^2, \ldots, s_r^2, 0, \ldots, 0$. Therefore, $AA^*u_i = s_i^2 u_i$ for $1 \le i \le r$; and $AA^*u_i = 0$ for $i > r$.

(b) As in (a), $Q^*A^*AQ = \text{diag}(s_1^2, \ldots, s_r^2, 0, \ldots, 0)$ proves analogous facts about the vectors v_j.

(3) Let $1 \le i \le r$. The vector u_i is the ith column of AQD. Thus

$$u_i = AQDe_i = AQ(s_i)^{-1}e_i = (s_i)^{-1}AQe_i = (s_i)^{-1}Av_i.$$

Using this and (2a), we obtain

$$(s_i)^{-1}A^*u_i = (s_i)^{-2}A^*Av_i = (s_i)^{-2}(s_i)^2 v_i = v_i.$$

(4) (a) For $1 \le i \le r$, $u_i = (s_i)^{-1}Av_i$ implies that $u_i \in R(A)$. The vectors $u_1, \ldots, u_r$ are orthonormal and $\dim(R(A)) - r$. Therefore, $\{u_1, \ldots, u_r\}$ is an orthonormal basis of $R(A)$.

(b) As in (a), $\{v_1, \ldots, v_r\}$ is an orthonormal basis of $R(A^*)$.

(c) Let $r < j \le n$. Now, $A^*Av_j = 0$ implies $v_j^* A^*Av_j = 0$. So, $\|Av_j\|^2 = 0$; or that $Av_j = 0$. Then $v_j \in N(A)$. But $\dim(N(A)) = n - r$. Therefore, the $n - r$ orthonormal vectors $v_{r+1}, \ldots, v_n$ form an orthonormal basis for $N(A)$.

(d) As in (c), $\{u_{r+1}, \ldots, u_m\}$ is an orthonormal basis for $N(A^*)$. ∎

Theorem 6.14(2) and (4) imply that the columns of P are eigenvectors of AA^*, and the columns of Q are eigenvectors of A^*A. Accordingly, the columns of P are called the *left singular vectors* of A; and the columns of Q are called the *right singular vectors* of A. Notice that computing both sets of left and right singular vectors independently will not serve the purpose since they may not satisfy the equations in Theorem 6.14(3).

Example 6.8 To determine the SVD of the matrix $A = \begin{bmatrix} 1 & 0 & 1 & 0 \\ 0 & 1 & 0 & 1 \end{bmatrix}$, we compute

$$AA^* = \begin{bmatrix} 2 & 0 \\ 0 & 2 \end{bmatrix}, \quad A^*A = \begin{bmatrix} 1 & 0 & 1 & 0 \\ 0 & 1 & 0 & 1 \\ 1 & 0 & 1 & 0 \\ 0 & 1 & 0 & 1 \end{bmatrix}.$$

Since AA^* is smaller, we compute its eigenvalues and eigenvectors. The eigenvalues of AA^* with multiplicities are 2, 2; thus the eigenvalues of A^*A are 2, 2, 0, 0. Choosing simpler eigenvectors of AA^*, we have

$$s_1^2 = 2, \ u_1 = e_1; \quad s_2^2 = 2, \ u_2 = e_2.$$

Here, u_1 and u_2 are the left singular vectors. The corresponding right singular vectors are:

$$v_1 = \frac{1}{s_1} A^* u_1 = \frac{1}{\sqrt{2}} \begin{bmatrix} 1 & 0 \\ 0 & 1 \\ 1 & 0 \\ 0 & 1 \end{bmatrix} \begin{bmatrix} 1 & 0 \end{bmatrix} = \frac{1}{\sqrt{2}} \begin{bmatrix} 1 \\ 0 \\ 1 \\ 0 \end{bmatrix}, \quad v_2 = \frac{1}{s_1} A^* u_2 = \frac{1}{\sqrt{2}} \begin{bmatrix} 0 \\ 1 \\ 0 \\ 1 \end{bmatrix}.$$

These eigenvectors are associated with the eigenvalues 2 and 2 of A^*A. We need the eigenvectors v_3 and v_4 associated with the remaining eigenvalues 0 and 0, which should also form an orthonormal set along with v_1 and v_2. Thus, we solve $A^*Ax = 0$. With $x = [a, b, c, d]^t$, the equations are

$$a + c = 0 = b + d.$$

Two linearly independent solutions of these equations are obtained by setting $a = 1$, $b = 0$, $c = -1$, $d = 0$ and $a = 0$, $b = 1$, $c = 0$, $d = -1$. The corresponding vectors are $w_3 = [1, 0 - 1, 0]^t$ and $w_4 = [0, 1, 0, -1]^t$. We find that $\{v_1, v_2, w_3, w_4\}$ is an orthogonal set. We then orthonormalize w_3 and w_4 to obtain

$$v_3 = \frac{w_3}{\|w_3\|} = \frac{1}{\sqrt{2}} \begin{bmatrix} 1 \\ 0 \\ -1 \\ 0 \end{bmatrix}, \quad v_4 = \frac{w_4}{\|w_4\|} = \frac{1}{\sqrt{2}} \begin{bmatrix} 0 \\ 1 \\ 0 \\ -1 \end{bmatrix}.$$

We set $P = [u_1 \ u_2]$, $Q = [v_1 \ v_2 \ v_3 \ v_4]$, and Σ as the 2×4 matrix with singular values $\sqrt{2}, \sqrt{2}$ on the 2×2 first block, and then other entries 0. We see that

$$A = \begin{bmatrix} 1 & 0 & 1 & 0 \\ 0 & 1 & 0 & 1 \end{bmatrix} = \begin{bmatrix} 1 & 0 \\ 0 & 1 \end{bmatrix} \begin{bmatrix} \sqrt{2} & 0 & 0 & 0 \\ 0 & \sqrt{2} & 0 & 0 \end{bmatrix} \frac{1}{\sqrt{2}} \begin{bmatrix} 1 & 0 & 1 & 0 \\ 0 & 1 & 0 & 1 \\ 1 & 0 & -1 & 0 \\ 0 & 1 & 0 & -1 \end{bmatrix} = P \Sigma Q^*. \ \square$$

In the product $P \Sigma Q^*$, there are possibly many zero rows that do not contribute to the end result. Thus some simplifications can be done in the SVD. Let $A \in \mathbb{C}^{m \times n}$ where $m \leq n$. Suppose $A = P \Sigma Q^*$ is an SVD of A. Let the ith column of Q be denoted by $v_i \in \mathbb{C}^{n \times 1}$. Write

$$P_1 = P, \ Q_1 = \begin{bmatrix} v_1 & \cdots & v_m \end{bmatrix} \in \mathbb{C}^{m \times n}, \ \Sigma_1 = \mathrm{diag}(s_1, \ldots, s_r, 0, \ldots, 0) \in \mathbb{C}^{m \times m}.$$

Notice that P_1 is unitary and the m columns of Q_1 are orthonormal. In block form, we have

$$Q = \begin{bmatrix} Q_1 & Q_3 \end{bmatrix}, \ \Sigma = \begin{bmatrix} \Sigma_1 & 0 \end{bmatrix}, \ Q_3 = \begin{bmatrix} v_{m+1} & \cdots & v_n \end{bmatrix}.$$

Then

$$A = P \Sigma Q^* = P_1 \begin{bmatrix} \Sigma_1 & 0 \end{bmatrix} \begin{bmatrix} Q_1^* \\ Q_3^* \end{bmatrix} = P_1 \Sigma_1 Q_1^* \quad \text{for } m \leq n. \tag{6.1}$$

Similarly, when $m \geq n$, we may curtail P accordingly. That is, suppose the ith column of P is denoted by $u_i \in \mathbb{C}^{m \times 1}$. Write

$$P_2 = \begin{bmatrix} u_1 & \cdots & u_n \end{bmatrix} \in \mathbb{C}^{m \times n}, \quad \Sigma_2 = \mathrm{diag}(s_1, \ldots, s_r, 0, \ldots, 0) \in \mathbb{C}^{n \times n}, \quad Q_2 = Q.$$

Here, the n columns of P_2 are orthonormal, and Q_2 is unitary. We write

$$P = \begin{bmatrix} P_2 & P_3 \end{bmatrix}, \quad \Sigma = \begin{bmatrix} \Sigma_2 \\ 0 \end{bmatrix}, \quad P_3 = \begin{bmatrix} u_{n+1} & \cdots & u_m \end{bmatrix}.$$

Then

$$A = P \, \Sigma \, Q^* = \begin{bmatrix} P_2 & P_3 \end{bmatrix} \begin{bmatrix} \Sigma_2 \\ 0 \end{bmatrix} Q_2^* = P_2 \Sigma_2 Q_2^* \quad \text{for } m \geq n. \tag{6.2}$$

The two forms of SVD in (6.1)–(6.2), one for $m \leq n$ and the other for $m \geq n$ are called the **thin SVD** of A. Of course, for $m = n$, both the thin SVDs coincide with the SVD. For a unified approach to the thin SVDs, take $k = \min\{m, n\}$. Then a matrix $A \in \mathbb{C}^{m \times n}$ of rank r can be written as the product

$$A = P \, \Sigma \, Q^*$$

where $P \in \mathbb{C}^{m \times k}$ and $Q \in \mathbb{C}^{n \times k}$ have orthonormal columns, and $\Sigma \in \mathbb{C}^{k \times k}$ is the diagonal matrix $\mathrm{diag}(s_1, \ldots, s_r, 0, \ldots, 0)$ with $s_1, \ldots, s_r$ being the positive singular values of A.

It is possible to simplify the thin SVDs further by deleting the zero rows. Observe that in the product $P^* A Q$, the first r columns of P and the first r columns of Q produce S; the other columns of P and of Q give the zero blocks. Thus taking

$$\tilde{P} = \begin{bmatrix} u_1 & \cdots & u_r \end{bmatrix}, \quad \tilde{Q} = \begin{bmatrix} v_1 & \cdots & v_r \end{bmatrix},$$

a simplified decomposition of A is given by

$$A = \tilde{P} \, S \, \tilde{Q}^*,$$

where $\tilde{P} \in \mathbb{C}^{m \times r}$ and $\tilde{Q} \in \mathbb{C}^{r \times n}$ are having orthonormal columns. Such a decomposition is called the **tight SVD** of the matrix A. In the tight SVD, Each of the matrices A, $\tilde{P}$, S and $\tilde{Q}^*$ is of rank r.

Write $B = \tilde{P} S$ and $C = S \tilde{Q}^*$ to obtain

$$A = B \, \tilde{Q}^* = \tilde{P} \, C,$$

where both $B \in \mathbb{C}^{m \times r}$ and $C \in \mathbb{C}^{r \times n}$ are of rank r. It shows that each $m \times n$ matrix of rank r can be written as a product of an $m \times r$ matrix of rank r and an $r \times n$ matrix, also of rank r. We recognize it as the *full rank factorization* of A.

Example 6.9 Obtain SVD, tight SVD, and a full rank factorization of

$$A = \begin{bmatrix} 2 & -1 \\ -2 & 1 \\ 4 & -2 \end{bmatrix}.$$

Here, $A^*A = \begin{bmatrix} 24 & -12 \\ -12 & 6 \end{bmatrix}$. It has eigenvalues 30 and 0. Thus $s_1 = \sqrt{30}$. Notice that AA^* is a 3×3 matrix with eigenvalues 30, 0 and 0. We see that $r = \text{rank}(A) = $ the number of positive singular values of $A = 1$.

For the eigenvalue 30, we solve the equation $A^*A[a, b]^t = 30[a, b]^t$, that is,

$$24a - 12b = 30a, \quad -12a + 6b = 30b.$$

It has the solution $a = -2$, $b = 1$. So, a unit eigenvector of A^*A corresponding to the eigenvalue 30 is $v_1 = \frac{1}{\sqrt{5}}[-2, 1]^t$.

For the eigenvalue 0, the equations are

$$24a - 12b = 0, \quad -12a + 6b = 0.$$

Thus a unit eigenvector orthogonal to v_1 is $v_2 = \frac{1}{\sqrt{5}}[1, 2]^t$. Then,

$$u_1 = \frac{1}{\sqrt{30}} A v_1 = \frac{1}{\sqrt{30}} \begin{bmatrix} 2 & -1 \\ -2 & 1 \\ 4 & -2 \end{bmatrix} \begin{bmatrix} -2/\sqrt{5} \\ 1/\sqrt{5} \end{bmatrix} = \frac{1}{\sqrt{6}} \begin{bmatrix} -1 \\ 1 \\ -2 \end{bmatrix}.$$

Notice that $\|u_1\| = 1$. We extend $\{u_1\}$ to an orthonormal basis of $\mathbb{C}^{3\times 1}$. It is

$$\left\{ u_1 = \frac{1}{\sqrt{6}} \begin{bmatrix} -1 \\ 1 \\ -2 \end{bmatrix}, \quad u_2 = \frac{1}{\sqrt{2}} \begin{bmatrix} 1 \\ 1 \\ 0 \end{bmatrix}, \quad u_3 = \frac{1}{\sqrt{3}} \begin{bmatrix} 1 \\ -1 \\ 1 \end{bmatrix} \right\}.$$

Next, we take u_1, u_2, u_3 as the columns of P and v_1, v_2 as the columns of Q to obtain the SVD of A as

$$\begin{bmatrix} 2 & -1 \\ -2 & 1 \\ 4 & -2 \end{bmatrix} = P\Sigma Q^* = \begin{bmatrix} -1/\sqrt{6} & 1/\sqrt{2} & 1/\sqrt{3} \\ 1/\sqrt{6} & 1/\sqrt{2} & -1/\sqrt{3} \\ -2/\sqrt{6} & 0 & 1/\sqrt{3} \end{bmatrix} \begin{bmatrix} \sqrt{30} & 0 \\ 0 & 0 \\ 0 & 0 \end{bmatrix} \begin{bmatrix} -2/\sqrt{5} & 1/\sqrt{5} \\ 1/\sqrt{5} & 2/\sqrt{5} \end{bmatrix}^*.$$

For the tight SVD, we construct $\tilde{P}$ with its r columns as the the first r columns of P, $\tilde{Q}$ with its r columns as the first r columns of Q, and S as the $r \times r$ block consisting of first r singular values of A as the diagonal entries. With $r = 1$, we thus have the tight SVD as

$$\begin{bmatrix} 2 & -1 \\ -2 & 1 \\ 4 & -2 \end{bmatrix} = \tilde{P}S\tilde{Q}^* = \begin{bmatrix} -1/\sqrt{6} \\ 1/\sqrt{6} \\ -2/\sqrt{6} \end{bmatrix} \left[\sqrt{30} \right] \begin{bmatrix} -2/\sqrt{5} \\ 1/\sqrt{5} \end{bmatrix}^*.$$

In the tight SVD, using associativity of matrix product, we get the full rank factorizations as

$$\begin{bmatrix} 2 & -1 \\ -2 & 1 \\ 4 & -2 \end{bmatrix} = \begin{bmatrix} -\sqrt{5} \\ \sqrt{5} \\ -2\sqrt{5} \end{bmatrix} \begin{bmatrix} -2/\sqrt{5} \\ 1/\sqrt{5} \end{bmatrix}^* = \begin{bmatrix} -1/\sqrt{6} \\ 1/\sqrt{6} \\ -2/\sqrt{6} \end{bmatrix} \begin{bmatrix} -2/\sqrt{6} \\ \sqrt{6} \end{bmatrix}^*.$$

It may be checked that the columns of Q are eigenvectors of AA^*.   □

A singular value decomposition of a matrix is not unique. For, orthonormal bases that comprise the columns of P and Q can always be chosen differently. For instance, by multiplying ± 1 to an already constructed orthonormal basis, we may obtain another.

Also, it is easy to see that when $A \in \mathbb{R}^{m \times n}$, the matrices P and Q can be chosen to have real entries.

Singular value decomposition is the most important result for scientists and engineers, perhaps, next to the theory of linear equations. It shows clearly the power of eigenvalues and eigenvectors. The SVD in the summation form, in Theorem 6.14(1), looks like

$$A = s_1 u_1 v_1^* + \cdots + s_r u_r v_r^*.$$

Each matrix $u_i v_i^*$ is of rank 1. This means that if we know the first r singular values of A and we know their corresponding left and right singular vectors, we know A completely. This is particularly useful when A is a very large matrix of low rank. No wonder, SVD is used in image processing, various compression algorithms, and in principal components analysis.

Let $A = P \Sigma Q^*$ be an SVD of $A \in \mathbb{C}^{m \times n}$, and let $x \in \mathbb{C}^{n \times 1}$ be a unit vector. Let $s_1 \geq \cdots \geq s_r$ be the positive singular values of A. Write $y = Q^*x = [\alpha_1, \ldots, \alpha_n]^t$. Then

$$\|y\|^2 = \sum_{i=1}^{n} |\alpha_i|^2 = \|Q^*x\|^2 = x^*QQ^*x = x^*x = \|x\|^2 = 1.$$

$$\|Ax\|^2 = \|P \Sigma Q^*x\|^2 = x^*Q \Sigma^* P^* P \Sigma Q^*x = x^*Q \Sigma^2 Q^*x$$

$$= y^* \Sigma^2 y = \sum_{j=1}^{r} s_j^2 |\alpha_j|^2 \leq s_1^2 \sum_{j=1}^{r} |\alpha_j|^2 \leq s_1^2.$$

Also, for $x = v_1$, $\|Av_1\|^2 = s_1^2 \|u_1\|^2 = s_1^2$. Therefore, we conclude that

$$s_1 = \max\{\|Ax\| : x \in \mathbb{C}^{n \times 1}, \ \|x\| = 1\}.$$

That is, the first singular value s_1 gives the maximum magnification that a vector experiences under the linear transformation A. Similarly, from above workout it follows that

$$\|Ax\|^2 = \sum_{j=1}^r s_j^2 |\alpha_j|^2 \geq s_r^2 \sum_{j=1}^r |\alpha_j|^2 \geq s_r^2.$$

With $x = v_r$, we have $\|Av_r\|^2 = s_r^2 \|u_r\|^2 = s_r^2$. Hence,

$$s_r = \min\{\|Ax\| : x \in \mathbb{C}^{n \times 1}, \ \|x\| = 1\}.$$

That is, the minimum positive magnification is given by the smallest positive singular value s_r.

Notice that if x is a unit vector, then $s_r \leq \|Ax\| \leq s_1$.

Exercises for Sect. 6.5

1. Let $A \in \mathbb{C}^{m \times n}$. Show that the positive singular values of A are also the positive singular values of A^*.
2. Compute the singular value decompositions of the following matrices:

   (a) $\begin{bmatrix} 2 & -2 \\ 1 & -1 \\ -1 & 1 \end{bmatrix}$ (b) $\begin{bmatrix} 2 & 1 & 2 \\ -2 & 1 & 2 \end{bmatrix}$ (c) $\begin{bmatrix} 1 & 2 & 2 \\ 2 & 0 & -5 \\ 3 & 0 & 0 \end{bmatrix}$

3. Show that the matrices $\begin{bmatrix} 1 & 0 \\ 1 & 1 \end{bmatrix}$ and $\begin{bmatrix} 2 & -1 \\ 1 & 0 \end{bmatrix}$ are similar but they have different singular values.
4. Prove that if $\lambda_1, \ldots, \lambda_n$ are the eigenvalues of an $n \times n$ hermitian matrix, then its singular values are $|\lambda_1|, \ldots, |\lambda_n|$.
5. Let $A \in \mathbb{C}^{n \times n}$ and let $\lambda \in \mathbb{C}$. Show that $A - \lambda I$ is invertible iff $A^* - \bar{\lambda} I$ is invertible.
6. Let $A \in \mathbb{F}^{n \times n}$. If A has eigenvalues $\lambda_1, \ldots, \lambda_n$ and singular values $s_1, \ldots, s_n$, then show that $|\lambda_1 \cdots \lambda_n| = s_1 \cdots s_n$.

6.6 Polar Decomposition

Square matrices behave like complex numbers in many ways. One such example is a powerful representation of square matrices using a stretch and a rotation. This mimics the polar representation of a complex number as $z = re^{i\theta}$, where r is a non-negative real number representing the stretch; and $e^{i\theta}$ is a rotation. Similarly, a square matrix can be written as a product of a positive semi-definite matrix, representing the stretch, and a unitary matrix representing the rotation. We slightly generalize it to any $m \times n$ matrix.

A hermitian matrix $P \in \mathbb{F}^{n \times n}$ is called **positive semidefinite** iff $x^*Px \geq 0$ for each $x \in \mathbb{F}^{n \times 1}$. We use such a matrix in the following matrix factorization.

Theorem 6.15 (Polar decomposition) *Let* $A \in \mathbb{C}^{m \times n}$. *Then there exist positive semi-definite matrices* $P \in \mathbb{C}^{m \times m}$, $Q \in \mathbb{C}^{n \times n}$, *and a matrix* $U \in \mathbb{C}^{m \times n}$ *such that*

$$A = PU = UQ,$$

where $P^2 = AA^*$, $Q^2 = A^*A$, *and* U *satisfies the following:*

(1) *If* $m = n$, *then the* $n \times n$ *matrix* U *is unitary.*
(2) *If* $m < n$, *then the rows of* U *are orthonormal.*
(3) *If* $m > n$, *then the columns of* U *are orthonormal.*

Proof Let $A \in \mathbb{C}^{m \times n}$ be a matrix of rank r with positive singular values $s_1 \geq \cdots \geq s_r$. Write $k = \min\{m, n\}$. Let $A = B \Sigma E^*$ be the thin SVD of A, where $B \in \mathbb{C}^{m \times k}$, $E \in \mathbb{C}^{n \times k}$ have orthonormal columns, and $\Sigma \in \mathbb{C}^{k \times k}$ has first r diagonal entries as $s_1, \ldots, s_r$, all other entries 0. Since $B^*B = E^*E = I$, we have

$$A = B \Sigma E^* = (B \Sigma B^*)(BE^*) = (BE^*)(E \Sigma E^*).$$

With $U = BE^*$, $P = B \Sigma B^*$ and $Q = E \Sigma E^*$, we obtain the polar decompositions $A = PU = UQ$. Moreover,

$$
\begin{aligned}
P^2 &= B \Sigma B^* B \Sigma B^* = B \Sigma E^* E \Sigma B^* = AA^*, \\
Q^2 &= E \Sigma E^* E \Sigma E^* = E \Sigma B^* B \Sigma E^* = A^*A.
\end{aligned}
$$

Notice that $\Sigma^* = \Sigma$. Thus $P^* = (B \Sigma B^*)^* = B \Sigma^* B^* = P$. That is, P is hermitian. Write $D = \mathrm{diag}(\sqrt{s_1}, \ldots, \sqrt{s_r}, 0, \ldots, 0) \in \mathbb{C}^{k \times k}$. Then $D^*D = \Sigma$. For each $x \in \mathbb{C}^{m \times 1}$,

$$x^*Px = x^*B \Sigma B^*x = x^*BD^*DB^*x = \|DB^*x\|^2 \geq 0.$$

Therefore, P is positive semi-definite. Similarly, Q is shown to be positive semi-definite.

(1) If $m = n$, then both B and E are unitary, i.e., $B^*B = BB^* = I = E^*E = EE^*$. Then $U^*U = (BE^*)^*BE^* = EB^*BE^* = EE^* = I$, and $UU^* = BE^*(BE^*)^* = BE^*EB^* = BB^* = I$. So, U is unitary.

(2) If $m < n$, then $k = m$; B is a square matrix with orthonormal columns, thus unitary; and $E^*E = I$. We have $UU^* = BE^*EB^* = BB^* = I$. Thus U has orthonormal rows.

(3) If $m > n$, then $k = n$. We see that E is unitary, and $B^*B = I$. Therefore, $U^*U = EB^*BE^* = EE^* = I$. That is, U has orthonormal columns. ∎

Recall that a thin SVD is obtained from an SVD of $A = P \Sigma Q^*$ by keeping the first k columns of the larger of the two matrices P and Q, and then restricting Σ to its first $k \times k$ block, where $k = \min\{m, n\}$. Thus, the polar decomposition of A may be constructed directly from the SVD. It is as follows:

If $A \in \mathbb{C}^{m \times n}$ has SVD as $A = BDE^*$, then $A = PU = UQ$, where

$$
\begin{aligned}
m = n: \quad & U = BE^*, \quad && P = BDB^*, \quad && Q = EDE^*. \\
m < n: \quad & U = BE_1^*, \quad && P = BD_1 B^*, \quad && Q = E_1 D_1 E_1^*. \\
m > n: \quad & U = B_1 E^*, \quad && P = B_1 D_2 B_1^*, \quad && Q = ED_2 E^*.
\end{aligned}
$$

Here, E_1 is constructed from E by taking its first m columns; D_1 is constructed from D by taking its first m columns; B_1 is constructed from B by taking its first n columns; and D_2 is constructed from D by taking its first n rows.

Example 6.10 Consider the matrix $A = \begin{bmatrix} 2 & -1 \\ -2 & 1 \\ 4 & -2 \end{bmatrix}$ of Example 6.9. We had obtained its SVD as $A = BDE^*$, where

$$
B = \begin{bmatrix} -1/\sqrt{6} & 1/\sqrt{2} & 1/\sqrt{3} \\ 1/\sqrt{6} & 1/\sqrt{2} & -1/\sqrt{3} \\ -2/\sqrt{6} & 0 & 1/\sqrt{3} \end{bmatrix}, \quad D = \begin{bmatrix} \sqrt{30} & 0 \\ 0 & 0 \\ 0 & 0 \end{bmatrix}, \quad E = \begin{bmatrix} -2/\sqrt{5} & 1/\sqrt{5} \\ 1/\sqrt{5} & 2/\sqrt{5} \end{bmatrix}.
$$

Here, $A \in \mathbb{C}^{3 \times 2}$. Thus, Theorem 6.15 (3) is applicable; see the discussion following the proof of the theorem. We construct the matrices B_1 by taking first two columns of B, and D_2 by taking first two rows of D, as in the following:

$$
B_1 = \begin{bmatrix} -1/\sqrt{6} & 1/\sqrt{2} \\ 1/\sqrt{6} & 1/\sqrt{2} \\ -2/\sqrt{6} & 0 \end{bmatrix}, \quad D_2 = \begin{bmatrix} \sqrt{30} & 0 \\ 0 & 0 \end{bmatrix}.
$$

Then

$$
U = B_1 E^* = \frac{1}{\sqrt{6}} \begin{bmatrix} -1 & \sqrt{3} \\ 1 & \sqrt{3} \\ -2 & 0 \end{bmatrix} \frac{1}{\sqrt{5}} \begin{bmatrix} -2 & 1 \\ 1 & 2 \end{bmatrix} = \frac{1}{\sqrt{30}} \begin{bmatrix} 2+\sqrt{3} & -1+2\sqrt{3} \\ -2+\sqrt{3} & 1+2\sqrt{3} \\ 4 & -2 \end{bmatrix},
$$

$$
P = B_1 D_2 B_1^* = \sqrt{5} \begin{bmatrix} -1 & 0 \\ 1 & 0 \\ -2 & 0 \end{bmatrix} \frac{1}{\sqrt{6}} \begin{bmatrix} -1 & 1 & -2 \\ \sqrt{3} & \sqrt{3} & 0 \end{bmatrix} = \frac{\sqrt{5}}{\sqrt{6}} \begin{bmatrix} 1 & -1 & 2 \\ -1 & 1 & -2 \\ 2 & -2 & 4 \end{bmatrix},
$$

$$
Q = ED_2 E^* = \sqrt{6} \begin{bmatrix} -2 & 0 \\ 1 & 0 \end{bmatrix} \frac{1}{\sqrt{5}} \begin{bmatrix} -2 & 1 \\ 1 & 2 \end{bmatrix} = \frac{\sqrt{6}}{\sqrt{5}} \begin{bmatrix} 4 & -2 \\ -2 & 1 \end{bmatrix}.
$$

As expected we find that

$$
PU = \frac{\sqrt{5}}{\sqrt{6}} \begin{bmatrix} 1 & -1 & 2 \\ -1 & 1 & -2 \\ 2 & -2 & 4 \end{bmatrix} \frac{1}{\sqrt{30}} \begin{bmatrix} 2+\sqrt{3} & -1+2\sqrt{3} \\ -2+\sqrt{3} & 1+2\sqrt{3} \\ 4 & -2 \end{bmatrix} = \begin{bmatrix} 2 & -1 \\ -2 & 1 \\ 4 & -2 \end{bmatrix} = A.
$$

$$
UQ = \frac{1}{\sqrt{30}} \begin{bmatrix} 2+\sqrt{3} & -1+2\sqrt{3} \\ -2+\sqrt{3} & 1+2\sqrt{3} \\ 4 & -2 \end{bmatrix} \frac{\sqrt{6}}{\sqrt{5}} \begin{bmatrix} 4 & -2 \\ -2 & 1 \end{bmatrix} = \begin{bmatrix} 2 & -1 \\ -2 & 1 \\ 4 & -2 \end{bmatrix} = A. \quad \square
$$

In $A = PU = UQ$, the matrices P and Q satisfy $P^2 = AA^*$ and $Q^2 = A^*A$. If $A \in \mathbb{C}^{m \times n}$, then $AA^* \in \mathbb{C}^{m \times m}$ and $A^*A \in \mathbb{C}^{n \times n}$ are hermitian matrices with eigenvalues as $s_1^2, \ldots, s_r^2, 0, \ldots, 0$. The diagonalization of AA^* yields m linearly independent eigenvectors associated with these eigenvalues. If these eigenvectors are taken as the columns (in that order) of a matrix C, then

$$AA^* = C \operatorname{diag}(s_1^2, \ldots, s_r^2, 0, \ldots, 0) \, C^*, \quad P = C \operatorname{diag}(s_1, \ldots, s_r, 0, \ldots, 0) \, C^*.$$

Similarly, the matrix Q is equal to $M \operatorname{diag}(s_1, \ldots, s_r, 0, \ldots, 0) \, M^*$, where M consists of orthonormal eigenvectors of A^*A corresponding to the eigenvalues $s_1^2, \ldots, s_r^2, 0, \ldots, 0$.

Now, using the matrix P as computed above, we can determine U so that $A = PU$. Similarly, the matrix U in $A = UQ$ can also be determined. In this approach, SVD is not used; however, the two instances of U may differ since they depend on the choices of orthonormal eigenvectors of AA^* and A^*A. If A is invertible, you would end up with the same U.

Exercises for Sect. 6.6

1. Let A be an upper triangular matrix with distinct diagonal entries. Show that there exists an upper triangular matrix that diagonalizes A.
2. Let $A = \begin{bmatrix} 1 & -1 \\ 0 & 1 \end{bmatrix}$. Determine whether A and A^2 are positive definite.
3. Determine the polar decompositions of the matrix A of Example 6.10 by diagonalizing AA^* and A^*A as mentioned in the text.
4. Let $A \in \mathbb{C}^{m \times n}$ with $m < n$. Prove that A can be written as $A = PU$, where $P \in \mathbb{C}^{m \times n}$, and $U \in \mathbb{C}^{n \times n}$ is unitary.
5. Let $A \in \mathbb{C}^{m \times n}$ with $m > n$. Prove that A can be written as $A = UQ$, where $U \in \mathbb{C}^{m \times m}$ is a unitary matrix and $Q \in \mathbb{C}^{m \times n}$.

6.7 Problems

1. Let P^*AP be the upper triangular matrix in Schur triangularization of $A \in \mathbb{F}^{n \times n}$. If A has n distinct eigenvalues, then show that there exists an upper triangular matrix Q such that PQ diagonalizes A.
2. If $A \in \mathbb{F}^{n \times n}$ is both diagonalizable and invertible, then how do you compute its inverse from its diagonalization?
3. Let $A \in \mathbb{F}^{n \times n}$ be diagonalizable with eigenvalues ± 1. Is $A^{-1} = A$?
4. How to diagonalize A^* if diagonalization of $A \in \mathbb{F}^{n \times n}$ is known?
5. Let $A \in \mathbb{R}^{n \times n}$ have real eigenvalues $\lambda_1 > \lambda_2 > \cdots > \lambda_n$ with corresponding eigenvectors $x_1, \ldots, x_n$. Let $x = \sum_{i=1}^{n} \alpha_i x_i$ for real numbers $\alpha_1, \ldots, \alpha_n$. Show the following:
   (a) $A^m x = \sum_{i=1}^{n} \alpha_i \lambda_i^m x_i$ (b) If $\lambda_1 = 1$, then $\lim_{m \to \infty} A^m x = \alpha_1 x_1$.

6. Prove that if a normal matrix has only real eigenvalues, then it is hermitian. Conclude that if a real normal matrix has only real eigenvalues, then it is real symmetric.

7. Let $A \in \mathbb{F}^{n \times n}$ have distinct eigenvalues $\lambda_1, \ldots, \lambda_k$. For $1 \leq j \leq k$, let the linearly independent eigenvectors associated with λ_j be $v_j^1, \ldots, v_j^{i_j}$. Prove that the set $\{v_1^1, \ldots, v_1^{i_1}, \ldots, v_k^1, \ldots, v_k^{i_k}\}$ is linearly independent. [Hint: See the proof of Theorem 5.3.]

8. Let $A = [a_{ij}] \in \mathbb{F}^{3 \times 3}$ be such that $a_{31} = 0$ and $a_{32} \neq 0$. Show that each eigenvalue of A has geometric multiplicity 1.

9. Suppose $A \in \mathbb{F}^{4 \times 4}$ has an eigenvalue λ with algebraic multiplicity 3, and rank$(A - \lambda I) = 1$. Is A diagonalizable?

10. Show that there exists only one $n \times n$ diagonalizable matrix with an eigenvalue λ of algebraic multiplicity n.

11. Show that a nonzero nilpotent matrix is never diagonalizable.
    [Hint: $A \neq 0$ but $A^m = 0$ for some $m \geq 2$.]

12. Let $A \in \mathbb{F}^{n \times n}$ and let $P^{-1}AP$ be a diagonal matrix. Show that the columns of P that are eigenvectors associated with nonzero eigenvalues of A form a basis for $R(A)$.

13. Let A be a diagonalizable matrix. Show that the number of nonzero eigenvalues of A is equal to rank(A).

14. Construct non-diagonalizable matrices A and B satisfying

    (a) rank(A) is equal to the number of nonzero eigenvalues of A;
    (b) rank(B) is not equal to the number of nonzero eigenvalues of B.

15. Let $x, y \in \mathbb{F}^{n \times 1}$ and let $A = xy^*$. Show the following:

    (a) A has an eigenvalue y^*x with an associated eigenvector x.
    (b) 0 is an eigenvalue of A with geometric multiplicity at least $n - 1$.
    (c) If $y^*x \neq 0$, then A is diagonalizable.

16. Using the Jordan form of a matrix show that a matrix A is diagonalizable iff for each eigenvalue of A, its geometric multiplicity is equal to its algebraic multiplicity.

17. Let $A \in \mathbb{C}^{n \times n}$ have an eigenvalue λ with algebraic multiplicity m. Prove that null$((A - \lambda I)^m) = m$.

18. Let λ be an eigenvalue of a matrix $A \in \mathbb{C}^{n \times n}$ having algebraic multiplicity m. Prove that for each $k \in \mathbb{N}$, if null$((A - \lambda I)^k) < m$, then null$((A - \lambda I))^k <$ null$((A - \lambda I)^{k+1}$.
    [Hint: Show that $N(A - \lambda I)^i \subseteq N(A - \lambda I)^{i+1}$. Then use Exercise 17.]

19. Let λ be an eigenvalue of a matrix A and let J be the Jordan form of A. Prove that the number of Jordan blocks with diagonal entry λ in J is the geometric multiplicity of λ.

20. Let A be a hermitian $n \times n$ matrix with eigenvalues $\lambda_1, \ldots, \lambda_n$. Show that there exist an orthonormal set $\{x_1, \ldots, x_n\}$ in $\mathbb{F}^{n \times 1}$ such that $x^*Ax = \sum_{i=1}^{n} \lambda_i |x^*x_i|^2$ for each $x \in \mathbb{F}^{n \times 1}$.

21. Let $A \in \mathbb{R}^{n \times n}$. Show that AA' and $A'A$ are similar matrices.
22. Let A and B be hermitian matrices of the same order. Are the following statements true?

    (a) All eigenvalues of AB are real.
    (b) All eigenvalues of ABA are real.

23. Let $n > 1$ and let $u \in \mathbb{F}^{n \times 1}$ be a unit vector. Let $H = I - 2uu^*$. Show the following:

    (a) H is both hermitian and unitary; thus $H^{-1} = H$.
    (b) If λ and μ are two distinct eigenvalues of H, then $|\lambda - \mu| = 2$.
    (c) $Hu = -u$.
    (d) The trace of H is $n - 2$.
    (e) If $v \neq 0$ and $v \perp u$, then $Hv = v$.
    (f) The eigenvalue 1 has algebraic multiplicity $n - 1$.

24. Let $A \in \mathbb{C}^{m \times n}$ be a matrix of rank r. Let $s_1 \geq \cdots \geq s_r$ be the positive singular values of A. Let $A = P \Sigma_1 Q^*$ be a singular value decomposition of A, with $S = \mathrm{diag}(s_1, \ldots, s_r)$, and $\Sigma_1 = \begin{bmatrix} S & 0 \\ 0 & 0 \end{bmatrix} \in \mathbb{C}^{m \times n}$. Define matrices $\Sigma_2 = \begin{bmatrix} S^{-1} & 0 \\ 0 & 0 \end{bmatrix} \in \mathbb{C}^{n \times m}$ and $A^{\dagger} = Q \Sigma_2 P^*$. Prove the following:

    (a) The matrix $A^{\dagger}$ satisfies the following properties:

    $$(AA^{\dagger})^* = AA^{\dagger}, \quad (A^{\dagger}A)^* = A^{\dagger}A, \quad AA^{\dagger}A = A, \quad A^{\dagger}AA^{\dagger} = A^{\dagger}.$$

    (b) There exists a unique matrix $A^{\dagger} \in \mathbb{F}^{n \times m}$ satisfying the four equations mentioned in (a). The matrix $A^{\dagger}$ is called the *generalized inverse* of A.
    (c) For any $b \in \mathbb{F}^{m \times 1}$, $A^{\dagger}b$ is the least squares solution of $Ax = b$.

25. Show that if s is a singular value of a matrix A, then there exists a nonzero vector x such that $\|Ax\| = s\|x\|$.
26. Let $A = P \Sigma Q^t$ be the SVD of a real $n \times n$ matrix. Let u_i be the ith column of P, and let v_i be the ith column of Q. Define the matrix B and the vectors x_i, y_i as follows:

    $$B = \begin{bmatrix} 0 & A^t \\ A & 0 \end{bmatrix}, \quad x_i = \begin{bmatrix} v_i \\ u_i \end{bmatrix}, \quad y_i = \begin{bmatrix} -v_i \\ u_i \end{bmatrix} \quad \text{for } 1 \leq i \leq n.$$

    Show that x_i and y_i are eigenvectors of B. How are the eigenvalues of B related to the singular values of A?
27. Derive the polar decomposition from the SVD. Also, derive singular value decomposition from the polar decomposition.
28. A *positive definite* matrix is a hermitian matrix such that for each $x \neq 0$, $x^*Ax > 0$. Show that a hermitian matrix is positive definite iff all its eigenvalues are positive.

29. Show that the square of a real symmetric invertible matrix is positive definite.
30. Show that if A is positive definite, then so is A^{-1}. Give an example of a 2×2 invertible matrix which is not positive definite.
31. Show that A^*A is positive semi-definite for any $A \in \mathbb{F}^{m \times n}$. Give an example of a matrix A where A^*A is not positive definite.
32. Show that if Q is unitary and A is positive definite, then QAQ^* is positive definite.
33. For a matrix $A \in \mathbb{F}^{n \times n}$, the *principal submatrices* are obtained by deleting its last r rows and last r columns for $r = 0, 1, \ldots, n - 1$. Show that all principal submatrices of a positive definite matrix are positive definite. Further, verify that all principal submatrices of $A = \begin{bmatrix} 1 & 1 & -3 \\ 1 & 1 & -3 \\ -3 & -3 & 5 \end{bmatrix}$ have non-negative determinants but A is not positive semi-definite.
34. Let A be a real symmetric matrix. Show that the following are equivalent:

    (a) A is positive definite.
    (b) All principal submatrices of A have positive determinant.
    (c) A can be reduced to an upper triangular form using only elementary row operations of Type3, where all pivots are positive.
    (d) $A = U^t U$, where U is upper triangular with positive diagonal entries.
    (e) $A = B^t B$ for some invertible matrix B.

35. Let A be a real symmetric positive definite $n \times n$ matrix. Show the following:

    (a) All diagonal entries of A are positive.
    (b) For any invertible $n \times n$ matrix P, $P^t A P$ is positive definite.
    (c) There exists an $n \times n$ orthogonal matrix Q such that $A = Q^t Q$.
    (d) There exist unique $n \times n$ matrices U and D where U is upper triangular with all diagonal entries 1, and D is a diagonal matrix with positive entries on the diagonal such that $A = U^t D U$.
    (e) *Cholesky factorization*: There exists a unique upper triangular matrix with positive diagonal entries such that $A = U^t U$.

Chapter 7
Norms of Matrices

7.1 Norms

Recall that the norm of a vector is the non-negative square root of the inner product of a vector with itself. Norms give an idea on the length of a vector. We wish to generalize on this theme so that we may be able to measure the length of a vector without resorting to an inner product. We keep the essential properties that are commonly associated with the length.

Let V be a subspace of $\mathbb{F}^n$. A **norm** on V is a function from V to $\mathbb{R}$ which we denote by $\| \cdot \|$, satisfying the following properties:

1. For each $v \in V$, $\|v\| \geq 0$.
2. For each $v \in V$, $\|v\| = 0$ iff $v = 0$.
3. For each $v \in V$ and for each $\alpha \in \mathbb{F}$, $\|\alpha v\| = |\alpha| \, \|v\|$.
4. For all $u, v \in V$, $\|u + v\| \leq \|u\| + \|v\|$.

Once a norm is defined on V, we call it a **normed linear space**. Though norms can be defined on any vector space, we require only subspaces of $\mathbb{F}^n$. In what follows, a *finite dimensional normed linear space* V will mean a subspace of some $\mathbb{F}^n$ in which a norm $\| \cdot \|$ has been defined.

Recall that Property (4) of a norm is called the *triangle inequality*. As we had seen earlier, in $\mathbb{R}^2$,

$$\|(a, b)\| = \left(|a|^2 + |b|^2\right)^{1/2} \quad \text{for } a, b \in \mathbb{R}$$

defines a norm. This norm comes from the usual inner product on $\mathbb{R}^2$. Some of the useful norms on $\mathbb{F}^n$ are discussed in the following example.

Example 7.1 Let $V = \mathbb{F}^n$. Let $v = (a_1, \ldots, a_n) \in \mathbb{F}^n$.

1. The function $\| \cdot \|_\infty : V \to \mathbb{R}$ given by $\|v\|_\infty = \max\{|a_1|, \ldots, |a_n|\}$ defines a norm on V. It is called the ∞-*norm* or the *Cartesian norm*.

2. The function $\| \cdot \|_1 : V \to \mathbb{R}$ given by $\|v\|_1 = |a_1| + \cdots + |a_n|$ defines a norm on V. It is called the 1-*norm* or the *taxicab norm*.

3. The function $\| \cdot \|_2 : V \to \mathbb{R}$ given by $\|v\|_2 = \left(|a_1|^2 + \cdots + |a_n|^2\right)^{1/2}$ is a norm. This norm is called the 2-*norm* or the *Euclidean norm*. In matrix form, it may be written as follows:

$$\text{for } v \in \mathbb{F}^{n \times 1}, \ \|v\|_2 = \sqrt{v^*v}\,; \quad \text{and} \quad \text{for } v \in \mathbb{F}^{1 \times n}, \ \|v\|_2 = \sqrt{vv^*}\,.$$

4. Let $p > 1$. Then, the function $\| \cdot \|_p : V \to \mathbb{R}$ defined by

$$\|v\|_p = \left(|a_1|^p + \cdots + |a_n|^p\right)^{1/p}$$

is a norm; it is called the *p-norm*.

$\square$

Consider the ∞-norm on $\mathbb{F}^2$. With $x = (1, 0)$, $y = (0, 1)$, we see that

$$\|(1, 0) + (0, 1)\|_\infty^2 + \|(1, 0) - (0, 1)\|_\infty^2 = (\max\{1, 1\})^2 + (\max\{1, 1\})^2 = 2.$$
$$2\left(\|(1, 0)\|_\infty^2 + \|(0, 1)\|_\infty^2\right) = 2\left((\max\{1, 0\})^2 + (\max\{0, 1\})^2\right) = 4.$$

Hence, there exist $x, y \in \mathbb{F}^n$ such that

$$\|x + y\|_\infty^2 + \|x - y\|_\infty^2 \neq 2\left(\|x\|_\infty^2 + \|y\|_\infty^2\right).$$

This violates the parallelogram law. Therefore, the ∞-norm does not come from an inner product. Similarly, it is easy to check that the 1-norm does not come from an inner product.

The normed linear space $\mathbb{F}^n$ with the Euclidean norm is different from $\mathbb{F}^n$ with the ∞-norm.

Notice that a norm behaves like the absolute value function in $\mathbb{R}$ or $\mathbb{C}$. With this analogy, we see that the *reverse triangle inequality* holds for norms also. It is as follows. Let $x, y \in V$, a normed linear space. Then,

$$\|x\| - \|y\| = \|(x - y) + y\| - \|y\| \leq \|x - y\| + \|y\| - \|y\| = \|x - y\|.$$

Similarly, $\|y\| - \|x\| \leq \|y - x\| = \|x - y\|$. Therefore,

$$\big| \|x\| - \|y\| \big| \leq \|x - y\| \quad \text{for all } x, y \in V.$$

This inequality becomes helpful in showing that the norm is a continuous functional. Recall that a *functional* on a subspace V of $\mathbb{F}^n$ is a function that maps vectors in V to the scalars in $\mathbb{F}$.

Any functional $f : V \to \mathbb{R}$ is **continuous at** $v \in V$ iff for each $\varepsilon > 0$, there exists a $\delta > 0$ such that for each $x \in V$, if $\|x - v\| < \delta$ then $|f(x) - f(v)| < \varepsilon$.

To show that $\| \cdot \| : V \to \mathbb{R}$ is continuous, let $\varepsilon > 0$. We take $\delta = \varepsilon$ and then verify the requirements. So, if $\|x - v\| < \varepsilon$, then $\big| \|x\| - \|v\| \big| \leq \|x - v\| < \varepsilon$. Therefore, the norm $\| \cdot \|$ is a continuous function.

Sometimes an estimate becomes easy by using a particular norm rather than the other. On $\mathbb{F}^n$, the following relations hold between the ∞-norm, 1-norm, and 2-norm:

$$\|v\|_\infty \leq \|v\|_2 \leq \|v\|_1, \quad \|v\|_1 \leq \sqrt{n}\, \|v\|_2 \leq n\, \|v\|_\infty.$$

You should be able to prove these facts on your own.

In fact, a generalization of the above inequalities exists. Its proof uses a fact from analysis that any continuous function from the closed unit sphere to $\mathbb{R}$ attains its minimum and maximum. If V is a normed linear space, then the *closed unit sphere* in V is defined as

$$S = \{v \in V : \|v\| = 1\}.$$

We use this fact in proving the following theorem.

Theorem 7.1 *Let V be a subspace of $\mathbb{F}^n$. Let $\| \cdot \|_a$ and $\| \cdot \|_b$ be two norms on V and let $v \in V$. Then, there exist positive constants α and β independent of v such that $\|v\|_a \leq \alpha \|v\|_b$ and $\|v\|_b \leq \beta \|v\|_a$.*

Proof Let $S_b = \{x \in V : \|x\|_b = 1\}$ be the unit sphere in V with respect to the norm $\| \cdot \|_b$. Since $\| \cdot \|_a$ is a continuous function, it attains a maximum on S_b. So, let

$$\alpha = \max \big\{ \|x\|_a : x \in S_b \big\}.$$

Then, for each $y \in S_b$, we have $\|y\|_a \leq \alpha$.

Let $v \in V$, $v \neq 0$. Take $y = v/\|v\|_b$. Now, $v \in S_b$. Then,

$$\|v\|_a = \big\| \|v\|_b y \big\|_a = \|v\|_b \|y\|_a \leq \alpha \|v\|_b.$$

Similarly, considering the continuous function $\| \cdot \|_b$ on the closed sphere $S_a = \{x \in V : \|x\|_a = 1\}$, we obtain the positive constant

$$\beta = \max \big\{ \|x\|_b : x \in S_a \big\}$$

so that for each $z \in S_a$, $\|z\|_b \leq \beta$. Then, with $z = v/\|v\|_a$, we have

$$\|v\|_b = \big\| \|v\|_a z \big\|_b = \|v\|_a \|z\|_b \leq \beta \|v\|_a.$$

If $v = 0$, then clearly both the inequalities hold with $\alpha = 1 = \beta$. ∎

Whenever the conclusion of Theorem 7.1 holds for two norms $\| \cdot \|_a$ and $\| \cdot \|_b$, we say that these two norms are **equivalent.** We thus see that on any subspace of $\mathbb{F}^n$, any two norms are equivalent. We will use the equivalence of norms later for defining a particular type of norms for matrices. We remark that on infinite dimensional normed linear spaces, any two norms need not be equivalent.

Exercises for Sect. 7.1

1. Show that the 1-norm on $\mathbb{F}^n$ does not come from an inner product.
2. Let $p \in \mathbb{N}$. Show that the p-norm is indeed a norm.
3. Let $v \in \mathbb{F}^n$. Show the following inequalities:
   (a) $\|v\|_\infty \le \|v\|_2 \le \|v\|_1$ (b) $\|v\|_1 \le \sqrt{n}\, \|v\|_2 \le n\, \|v\|_\infty$

7.2 Matrix Norms

Norms provide a way to quantify the vectors. It is easy to verify that the addition of $m \times n$ matrices and multiplying a matrix with a scalar satisfy the properties required of a vector space. Thus, $\mathbb{F}^{m \times n}$ is a vector space. We can view $\mathbb{F}^{m \times n}$ as a normed linear space by providing a norm on it.

Example 7.2 The following functions $\| \cdot \| : \mathbb{F}^{m \times n} \to \mathbb{R}$ define norms on $\mathbb{F}^{m \times n}$. Let $A = [a_{ij}] \in \mathbb{F}^{m \times n}$.

1. $\|A\|_c = \max \big\{ |a_{ij}| : 1 \le i \le m, \ 1 \le j \le n \big\}.$
   It is called the *Cartesian norm* on matrices.
2. $\|A\|_t = \sum_{i=1}^m \sum_{j=1}^n |a_{ij}|.$
   It is called the *taxicab norm* on matrices.
3. $\|A\|_F = \big(\sum_{i=1}^m \sum_{j=1}^n |a_{ij}|^2 \big)^{1/2} = (\mathrm{tr}(A^*A))^{1/2}.$
   It is called the *Frobenius norm* on matrices.

$\square$

Notice that the names of the norms in Example 7.2 follow the pattern in Example 7.1. However, the notation uses the subscripts c, t, F instead of ∞, 1, 2. The reason is that we are reserving the latter notation for some other matrix norms, which we will discuss soon.

For matrices, it will be especially useful to have such a norm which satisfies $\|Av\| \le \|A\| \, \|v\|$. It may quite well happen that an arbitrary vector norm and an arbitrary matrix norm may not satisfy this property. Thus, given a vector norm, we require to define a corresponding matrix norm so that such a property is satisfied. In order to do that, we will first prove a result.

Theorem 7.2 *Let $A \in \mathbb{F}^{m \times n}$. Let $\| \cdot \|_n$ and $\| \cdot \|_m$ be norms on $\mathbb{F}^{n \times 1}$ and $\mathbb{F}^{m \times 1}$, respectively. Then,* $\left\{ \dfrac{\|Av\|_m}{\|v\|_n} : v \in \mathbb{F}^{n \times 1}, \ v \ne 0 \right\}$ *is a bounded subset of $\mathbb{R}$.*

Proof Clearly, 0 is a lower bound for the given set. We need to show that the given set has an upper bound.

Let $\{e_1, \ldots, e_n\}$ be the standard basis of $\mathbb{F}^{n \times 1}$. For each i, $\|Ae_i\|_m$ is a real number. Write $\alpha = \sum_{i=1}^n \|Ae_i\|_m$.

Let $v \in \mathbb{F}^{n \times 1}$, $v \neq 0$. We have unique scalars $\beta_1, \ldots, \beta_n$ not all zero such that $v = \sum_{i=1}^{n} \beta_i e_i$. Then, $\|v\|_\infty = \max\{|\beta_i| : 1 \leq i \leq n\}$. And,

$$\|Av\|_m = \left\| A\left(\sum_{i=1}^{n} \beta_i e_i\right) \right\|_m = \left\| \sum_{i=1}^{n} \beta_i A e_i \right\|_m = \sum_{i=1}^{n} |\beta_i| \, \|Ae_i\|_m$$

$$\leq \|v\|_\infty \sum_{i=1}^{n} \|Ae_i\|_m = \alpha \|v\|_\infty.$$

Consider the norms $\|\cdot\|_\infty$ and $\|\cdot\|_n$ on $\mathbb{F}^{n \times 1}$. Due to Theorem 7.1, there exists a positive constant γ such that $\|v\|_\infty \leq \gamma \|v\|_n$; the constant γ does not depend on particular vector v. Then, it follows that $\|Av\|_m \leq \alpha\gamma \|v\|_n$. That is, for each nonzero vector $v \in \mathbb{F}^{n \times 1}$,

$$\frac{\|Av\|_m}{\|v\|_n} \leq \alpha\gamma.$$

Therefore, the given set is bounded above by $\alpha\gamma$. Also, the set is bounded below by 0. ∎

The axiom of completeness of $\mathbb{R}$ asserts that every nonempty bounded subset of $\mathbb{R}$ has a least upper bound (lub) and a greatest lower bound (glb) in $\mathbb{R}$. Thus, the least upper bound of the set in Theorem 7.2 is a real number. Using this we define a type of norm on matrices.

Let $\|\cdot\|_m$ and $\|\cdot\|_n$ be norms on $\mathbb{F}^{m \times 1}$ and $\mathbb{F}^{n \times 1}$, respectively. The norm on $\mathbb{F}^{m \times n}$ given by

$$\|A\|_{m,n} = \mathrm{lub}\left\{ \frac{\|Av\|_m}{\|v\|_n} : v \in \mathbb{F}^{n \times 1},\ v \neq 0 \right\} \quad \text{for } A \in \mathbb{F}^{m \times n}$$

is called the **matrix norm induced by the vector norms** $\|\cdot\|_m$ and $\|\cdot\|_n$.

We use the phrase *induced norm* for matrices, for short. Verify that the induced norm is, in fact, a norm on $\mathbb{F}^{m \times n}$.

For square matrices, the induced norm takes an alternate form. For, suppose $\|\cdot\|$ is a norm on $\mathbb{F}^{n \times 1}$. The induced norm on $\mathbb{F}^{n \times n}$ is then given by

$$\|A\| = \mathrm{lub}\left\{ \frac{\|Av\|}{\|v\|} : v \in \mathbb{F}^{n \times 1},\ v \neq 0 \right\} = \mathrm{lub}\left\{ \|Ax\| : x \in \mathbb{F}^{n \times 1},\ \|x\| = 1 \right\}.$$

The induced norms on $\mathbb{F}^{n \times n}$ satisfy the desired properties with respect to the product of a vector with a matrix and also that of a matrix with another.

Theorem 7.3 *Let* $\|\cdot\|_m$, $\|\cdot\|_k$, *and* $\|\cdot\|_n$ *be norms on* $\mathbb{F}^{m \times 1}$, $\mathbb{F}^{k \times 1}$, *and* $\mathbb{F}^{n \times 1}$, *respectively. Let* $\|\cdot\|_{m,k}$ *and* $\|\cdot\|_{k,n}$ *be the corresponding induced norms on* $\mathbb{F}^{m \times k}$ *and* $\mathbb{F}^{k \times n}$, *respectively. Let* $A \in \mathbb{F}^{m \times k}$, $B \in \mathbb{F}^{k \times n}$, *and let* $v \in \mathbb{F}^{k \times 1}$. *Then,*

$$\|Av\|_m \leq \|A\|_{m,k} \|v\|_k \quad \text{and} \quad \|AB\|_{m,n} \leq \|A\|_{m,k} \|B\|_{k,n}.$$

Proof To keep the notation simple, let us write all the norms involved as $\| \cdot \|$; the subscripts may be supplied appropriately.

If $v = 0$, then $\|Av\| = 0 = \|v\|$. If $v \neq 0$, then

$$\frac{\|Av\|}{\|v\|} \leq \text{lub} \left\{ \frac{\|Av\|}{\|v\|} : v \in \mathbb{F}^{k \times 1}, \ v \neq 0 \right\} = \|A\|.$$

Hence, $\|Av\| \leq \|A\| \, \|v\|$.

For the second inequality, first, suppose that $Bx \neq 0$ for any $x \neq 0$ in $\mathbb{F}^{n \times 1}$. Then,

$$
\begin{aligned}
\|AB\| &= \text{lub} \left\{ \frac{\|ABx\|}{\|x\|} : x \in \mathbb{F}^{n \times 1}, \ x \neq 0 \right\} \\
&= \text{lub} \left\{ \frac{\|ABx\|}{\|Bx\|} \frac{\|Bx\|}{\|x\|} : x \in \mathbb{F}^{n \times 1}, \ x \neq 0, \ Bx \neq 0 \right\} \\
&\leq \text{lub} \left\{ \frac{\|ABx\|}{\|Bx\|} : x \in \mathbb{F}^{n \times 1}, \ Bx \neq 0 \right\} \text{lub} \left\{ \frac{\|Bx\|}{\|x\|} : x \in \mathbb{F}^{n \times 1}, \ x \neq 0 \right\} \\
&= \|A\| \, \|B\|.
\end{aligned}
$$

Next, if $Bx = 0$ for some $x \neq 0$, then $ABx = 0$. Thus, in the first line of the above calculation, restricting to the set $\{x : Bx \neq 0\}$ will not change the least upper bound. Further, if $Bx = 0$ for each x, then $\|B\| = 0 = \|AB\|$. ∎

A matrix norm that satisfies the property $\|AB\| \leq \|A\| \, \|B\|$ for all matrices for which the product AB is well defined is called a *sub-multiplicative norm*. Thus, the induced norm on matrices is sub-multiplicative.

Example 7.3 Define the function $\| \cdot \|_\infty : \mathbb{F}^{m \times n} \to \mathbb{R}$ by

$$\|A\|_\infty = \max \left\{ (|a_{i1}| + |a_{i2}| + \cdots + |a_{in}|) : 1 \leq i \leq m \right\} \quad \text{for } A = [a_{ij}] \in \mathbb{F}^{m \times n}.$$

Then, $\| \cdot \|_\infty$ is a norm on $\mathbb{F}^{m \times n}$. It is called the *maximum absolute row sum norm*, or the *row sum norm*, for short.

The row sum norm on matrices is induced by the ∞-norm on vectors. To see this, suppose that

$\| \cdot \|_\infty$ is the ∞-norm on the spaces $\mathbb{F}^{n \times 1}$ and $\mathbb{F}^{m \times 1}$,

$\| \cdot \|$ is the norm on matrices induced by the vector norm $\| \cdot \|_\infty$,

$\| \cdot \|_\infty$ is the row sum norm on matrices, as defined above,

$A = [a_{ij}] \in \mathbb{F}^{m \times n}$, and $v = (b_1, \ldots, b_n)^t \in \mathbb{F}^{n \times 1}$, $v \neq 0$.

Then (Write down in full and multiply.)

$$\|Av\|_\infty = \max\left\{\left|\sum_{j=1}^n a_{1j}b_j\right|, \ldots, \left|\sum_{j=1}^n a_{mj}b_j\right|\right\}$$

$$\leq \max\left\{\left|\sum_{j=1}^n a_{1j}\right||b_j|, \ldots, \left|\sum_{j=1}^n a_{mj}\right||b_j|\right\}$$

$$\leq \max\left\{\left|\sum_{j=1}^n a_{1j}\right|, \ldots, \left|\sum_{j=1}^n a_{mj}\right|\right\}\|v\|_\infty$$

$$\leq \|A\|_\infty \|v\|_\infty.$$

That is, for each nonzero vector $v \in \mathbb{F}^{n\times 1}$, $\dfrac{\|Av\|_\infty}{\|v\|_\infty} \leq \|A\|_\infty$. Then,

$$\|A\| = \text{lub}\left\{\frac{\|Av\|_\infty}{\|v\|_\infty} : v \in \mathbb{F}^{n\times 1},\ v \neq 0\right\} \leq \|A\|_\infty. \qquad (7.1)$$

To show that $\|A\| = \|A\|_\infty$, we construct a vector $v \in \mathbb{F}^{n\times 1}$ such that $\dfrac{\|Av\|_\infty}{\|v\|_\infty} = \|A\|_\infty$. If $A = 0$, then clearly, $\dfrac{\|Ae_1\|_\infty}{\|e_1\|_\infty} = 0 = \|A\|_\infty$. So, suppose $A \neq 0$. Now, the maximum of the sums of absolute values of entries in any row of A occurs at some row. Choose one such row index, say, k. Take the vector $u = (c_1, \ldots, c_n)^t$, where

$$c_j = \begin{cases} a_{kj}/|a_{kj}| & \text{if } a_{kj} \neq 0 \\ 0 & \text{if } a_{kj} = 0. \end{cases}$$

Then, $|c_j| = 0$ when $a_{kj} = 0$; otherwise, $|c_j| = 1$. Notice that since $A \neq 0$, there exists at least one j such that $c_j = 1$. Then, $\|u\|_\infty = 1$ and $\|Au\|_\infty = \|A\|_\infty$.

Using $\|A\|$ as the lub of the set as written in (7.1), we have

$$\|A\| \geq \frac{\|Au\|_\infty}{\|u\|_\infty} = \|Au\|_\infty = \|A\|_\infty.$$

Therefore, with (7.1), we conclude that $\|A\| = \|A\|_\infty$. $\qquad\square$

Example 7.4 Define the function $\|\cdot\|_1 : \mathbb{F}^{m\times n} \to \mathbb{R}$ by

$$\|A\|_1 = \max\left\{(|a_{1j}| + |a_{2j}| + \cdots + |a_{mj}|) : 1 \leq j \leq n\right\} \text{ for } A = [a_{ij}] \in \mathbb{F}^{m\times n}.$$

Then, $\|\cdot\|_1$ is a norm on $\mathbb{F}^{m\times n}$. It is called the *maximum absolute column sum norm* or the *column sum norm*, for short.

The column sum norm is induced by the 1-norm of vectors. To see this, suppose $A = [a_{ij}] \in \mathbb{F}^{m\times n}$ and $v = (b_1, \ldots, b_n)^t \in \mathbb{F}^{n\times 1}$. Denote by $A_1, \ldots, A_n$, the columns of A. Write the matrix norm induced by the vector norm $\|\cdot\|_1$ as $\|\cdot\|$. Using the triangle inequality, we obtain

$$\|Av\|_1 = \left\| \sum_{j=1}^{n} b_j A_j \right\|_1 \leq \sum_{j=1}^{n} \|b_j A_j\|_1$$

$$= \sum_{j=1}^{n} |b_j| \|A_j\|_1 \leq \max \left\{ \|A_j\|_1 : 1 \leq j \leq n \right\} \left(\sum_{j=1}^{n} |b_j| \right)$$

$$= \max \left\{ \|A_j\|_1 : 1 \leq j \leq n \right\} \|v\|_1 = \|A\|_1 \|v\|_1.$$

That is, for each nonzero vector $v \in \mathbb{F}^{n \times 1}$,

$$\frac{\|Av\|_1}{\|v\|_1} \leq \max \left\{ \|A_j\|_1 : 1 \leq j \leq n \right\} = \|A\|_1.$$

It shows that $\|A\| \leq \|A\|_1$.

For showing equality, suppose the maximum of the sums of absolute values of entries in any column occurs at the kth column. Again, this k may not be unique; but we choose one among the columns and fix such a column index as k. That is, $\|A_k\|_1 = \|A\|_1$. Now, $\|e_k\|_1 = 1$ and

$$\|A\| \geq \frac{\|Ae_k\|_1}{\|e_k\|_1} = \|A_k\|_1 = \|A\|_1.$$

Therefore, $\|A\| = \|A\|_1$.   □

Example 7.5 Let $A \in \mathbb{F}^{m \times n}$. Let $\| \cdot \|_2$ denote the matrix norm on $\mathbb{F}^{m \times n}$ induced by the Euclidean norm or the 2-norm $\| \cdot \|_2$ on vectors. Then,

$$\|A\|_2^2 = \mathrm{lub} \left\{ \frac{\|Av\|_2^2}{\|v\|_2^2} : v \in \mathbb{F}^{n \times 1}, \ v \neq 0 \right\} = \mathrm{lub} \left\{ \frac{v^* A^* A v}{v^* v} : v \in \mathbb{F}^{n \times 1}, \ v \neq 0 \right\}.$$

The matrix $A^* A$ is hermitian. Due to the spectral theorem, it is unitarily diagonalizable. Its eigenvalues can be ordered as $s_1^2 \geq \cdots \geq s_n^2 \geq 0$. So, let $B = \{v_1, \ldots, v_n\}$ be an orthonormal basis of $\mathbb{F}^{n \times 1}$, where v_j is an eigenvector of $A^* A$ corresponding to the eigenvalue s_j^2. Let $v \in \mathbb{F}^{n \times 1}$, $v \neq 0$. We have scalars α_j such that $v = \sum_{j=1}^{n} \alpha_j v_j$. Using orthonormality of v_js, we see that

$$v^* v = \sum_{i=1}^{n} \sum_{j=1}^{n} \overline{\alpha}_i \alpha_j v_i^* v_j = \sum_{i=1}^{n} |\alpha_i|^2.$$

$$v^* A^* A v = \left(\sum_{i=1}^{n} \overline{\alpha}_i v_i^* \right) \left(\sum_{j=1}^{n} s_j^2 \alpha_j v_j \right) = \sum_{i=1}^{n} \sum_{j=1}^{n} \overline{\alpha}_i \alpha_j s_j^2 v_i^* v_j = \sum_{i=1}^{n} |\alpha_i|^2 s_i^2.$$

$$s_1^2 - \frac{v^* A^* A v}{v^* v} = s_1^2 - \frac{\sum_{i=1}^{n} |\alpha_i|^2 s_i^2}{\sum_{i=1}^{n} |\alpha_i|^2} = \frac{\sum_{i=1}^{n} |\alpha_i|^2 (s_1^2 - s_i^2)}{\sum_{i=1}^{n} |\alpha_i|^2} \geq 0.$$

This is true for all $v \neq 0$. In particular, for $v = v_1$, we have $\alpha_1 = 1$ and $\alpha_i = 0$ for each $i > 1$. We thus obtain

$$s_1^2 - \frac{v_1^* A^* A v_1}{v_1^* v_1} = \frac{v_1^* s_1^2 v_1}{v_1^* v_1} = 0.$$

Therefore,

$$s_1^2 = \mathrm{lub} \left\{ \frac{v^* A^* A v}{v^* v} : v \in \mathbb{F}^{n \times 1}, \ v \neq 0 \right\}.$$

It shows that $\|A\|_2 = s_1$ is the required induced norm, where s_1 is the largest singular value of A. This induced norm is called the 2-norm and also the *spectral norm* on matrices. □

In general, the quotient $\rho_A(v) = \dfrac{v^* A v}{v^* v}$ is called the **Rayleigh quotient** of the matrix A with the nonzero vector v. The Rayleigh quotients $\rho_A(v_j)$ for eigenvectors v_j give the corresponding eigenvalues of A. It comes of help in computing an eigenvalue if the associated eigenvector is known. It can be shown that the Rayleigh quotient of a hermitian matrix with any nonzero vector is a real number. Moreover, if a hermitian matrix A has eigenvalues $\lambda_1 \geq \cdots \geq \lambda_n$, then $\lambda_1 \geq \rho_A(v) \geq \lambda_n$ for any nonzero vector v.

We see that the induced norms on $\mathbb{F}^{m \times n}$ with respect to the norms $\| \cdot \|_\infty$, $\| \cdot \|_1$, and $\| \cdot \|_2$ on both $\mathbb{F}^{n \times 1}$ and $\mathbb{F}^{m \times 1}$ are the row sum norm $\| \cdot \|_\infty$, the column sum norm $\| \cdot \|_1$, and the spectral norm $\| \cdot \|_2$, respectively.

If $\| \cdot \|$ is any induced norm on $\mathbb{F}^{n \times n}$, then $\dfrac{\|I(v)\|}{\|v\|} = 1$ says that $\|I\| = 1$.

For $n > 1$, the taxicab norm $\|I\|_t = n$; so, it is not an induced norm.

The Cartesian norm is not sub-multiplicative. For instance, take A and B as the $n \times n$ matrix with each of its entries equal to 1. Then, $\|A\|_c = \|B\|_c = 1$ and $\|AB\|_c = n$. Thus, it is not an induced norm for $n > 1$.

If $n > 1$, then the Frobenius norm of the identity is $\|I\|_F = \sqrt{n} > 1$. Thus, the Frobenius norm is not induced by any vector norm. However, it satisfies the sub-multiplicative property. For, Cauchy–Swartz inequality implies that

$$\|AB\|_F^2 = \sum_{i=1}^n \sum_{j=1}^n \left| \sum_{k=1}^n a_{ik} b_{kj} \right|^2 \leq \sum_{i=1}^n \sum_{j=1}^n \left(\sum_{k=1}^n |a_{ik}|^2 \right) \left(\sum_{k=1}^n |b_{kj}|^2 \right)$$

$$= \left(\sum_{i=1}^n \sum_{k=1}^n |a_{ik}|^2 \right) \left(\sum_{j=1}^n \sum_{k=1}^n |b_{kj}|^2 \right) = \|A\|_F^2 \|B\|_F^2.$$

The spectral norm and the Frobenius norms are mostly used in applications since both of them satisfy the sub-multiplicative property. The other norms are sometimes used for deriving easy estimates due to their computational simplicity.

Exercises for Sect. 7.2

1. Let $x \in \mathbb{R}^2$. Show that $\|x\|_2 = 1$ iff $x = (\cos\theta, \ \sin\theta)$ for some $\theta \in \mathbb{R}$.
2. Prove that the induced norm as defined in the text is indeed a norm. That is, it satisfies the required four properties of a norm.
3. Let $\|\cdot\|$ be a norm on $\mathbb{F}^{n \times 1}$. Let $A \in \mathbb{F}^{n \times n}$. Prove that

$$\text{lub} \left\{ \frac{\|Av\|}{\|v\|} : v \in \mathbb{F}^{n \times 1}, \ v \neq 0 \right\} = \text{lub} \left\{ \|Ax\| : x \in \mathbb{F}^{n \times 1}, \ \|x\| = 1 \right\}.$$

7.3 Contraction Mapping

We wish to discuss the use of norms in a special circumstance. Suppose it is required to solve the equation $x^2 - x - 1 = 0$ numerically, given that it has a root between 1 and 2. (Pretend that we do not know the formula for solving a quadratic equation.) We may then rewrite the equation as $x = \sqrt{1+x}$. As an initial guess, we take $x_0 = 1$. Using the iteration $x_{n+1} = \sqrt{1 + x_n}$, we find that

$$x_0 = 1, \ x_1 = \sqrt{2}, \ x_2 = \sqrt{1 + \sqrt{2}}, \ x_3 = \sqrt{1 + \sqrt{1 + \sqrt{2}}}, \ldots.$$

We compute the absolute difference between the successive approximants:

$$|x_1 - x_0| = \sqrt{2} - 1, \quad |x_2 - x_1| = \sqrt{1 + \sqrt{2}} - \sqrt{2},$$

$$|x_3 - x_2| = \sqrt{1 + \sqrt{1 + \sqrt{2}}} - \sqrt{1 + \sqrt{2}}, \ldots$$

Those seem to decrease. So, we conjecture that this iteration can be used to approximate the root of the equation that lies between 1 and 2. Our intuition relies on the fact that if the sequence $\{x_n\}$ converges to a real number x, then the limit x will satisfy the equation $x = \sqrt{1+x}$.

Moreover, we require the successive approximants to come closer to the root, or at least, the difference between the successive approximants decreases to 0. This requirement may put some restrictions on the function we use for iteration, such as $f(x) = \sqrt{1+x}$ above.

Let S be a nonempty subset of a normed linear space V. Let $f : S \to S$ be any function. We say that $v \in S$ is a **fixed point** of f iff $f(v) = v$. The function $f : S \to S$ is called a **contraction mapping** or a **contraction** iff there exists a positive number $c < 1$ such that

$$\|f(u) - f(v)\| \leq c \|u - v\| \quad \text{for all } u, v \in S.$$

Example 7.6 1. Let $V = \mathbb{R}$. Let S be the closed interval $\{x \in \mathbb{R} : 1 \leq x \leq 2\}$.
Define $f : S \to \mathbb{R}$ by $f(x) = \sqrt{1+x}$.
If $1 \leq x \leq 2$, then $\sqrt{2} \leq f(x) \leq \sqrt{3}$. That is, $f : S \to S$. If a is a fixed point of
f, then it satisfies $a = \sqrt{1+a}$ or $a^2 - a - 1 = 0$.
Now, the derivative of f is continuous on S, and it satisfies

$$|f'(x)| = \left| \frac{1}{2\sqrt{1+x}} \right| \leq \frac{1}{2\sqrt{3}} < 1.$$

Then, for all $x, y \in S$,

$$|f(x) - f(y)| \leq \left(\max \{|f'(x)| : x \in S\} \right)|x - y| \leq \frac{1}{2\sqrt{3}} |x - y|.$$

Therefore, f is a contraction.
2. Let $V = \mathbb{R}$, and $S = \mathbb{R}$ also. Define $f : S \to \mathbb{R}$ by $f(x) = x^2 - 1$. Its fixed point
a satisfies $a = a^2 - 1$ or $a^2 - a - 1 = 0$.
For $x = 2$, $y = 3$, $|x - y| = 1$, $|f(x) - f(y)| = |2^2 - 1 - (3^2 - 1)| = 5$. Thus,

$$|f(x) - f(y)| \not\leq c\,|x - y| \quad \text{for any } c < 1.$$

Hence, f is not a contraction.
3. Let $V = \mathbb{R}^{n \times 1}$ for a fixed $n > 1$. Let $S = V$. We use the 2-norm $\|v\|_2$ on V.
Define $f : S \to S$ by

$$f(v) = Av, \quad A = \text{diag}\left(0, \tfrac{1}{n}, \tfrac{1}{n} \ldots, \tfrac{1}{n}\right), \quad \text{for } v \in S.$$

Then, $v = 0$ is a fixed point of f.
Notice that $\|A\|_2 = $ the largest singular value of $A = \tfrac{1}{n} < 1$. Then,

$$\|f(u) - f(v)\|_2 = \|Au - Av\|_2 \leq \|A\|_2 \|u - v\|_2 = \tfrac{1}{n}\|u - v\| \quad \text{for } u, v \in S.$$

Hence, f is a contraction.
4. With $S = V = \mathbb{R}^{n \times 1}$ and $f(v) = v$ defined on S, we see that each vector $v \in \mathbb{R}^{n \times 1}$
is a fixed point of f.
For any norm on V, we have $\|f(u) - f(v)\| = \|u - v\|$. Therefore, f is not a
contraction.

$\square$

We will show that a fixed point can be approximated by using a contraction in an
iterative way. For a contraction $f : S \to S$, we will use the *fixed-point iteration*

$$x_0, \quad x_{n+1} = f(x_n) \quad \text{for } n \geq 0$$

Here, $x_0 \in S$ is chosen initially, and we say that the *iteration function* is f. This defines a sequence $\{x_n\}_{n=0}^{\infty}$ in S. We require this sequence to converge to a point in S. In this regard, we quote some relevant notions and facts from analysis.

Let S be a subset of a normed linear space V. We say that a sequence $\{y_n\}_{n=0}^{\infty}$ of vectors y_n is in S iff $y_n \in S$ for each n.

A sequence $\{y_n\}_{n=0}^{\infty}$ in S is said to **converge** to a vector $y \in V$ iff for each $\varepsilon > 0$, there exists a natural number N such that if $m > N$, then $\|y_m - y\| < \varepsilon$.

In such a case, the vector $y \in V$ is called a **limit** of the (convergent) sequence. If for all convergent sequences the corresponding limit vector happens to be in S, then S is said to be a **closed subset** of V.

Further, a sequence $\{y_n\}_{n=0}^{\infty}$ in S is called a **Cauchy sequence** iff for each $\varepsilon > 0$, there exists a natural number N such that if $m > N$ and $k > 0$, then $\|y_{m+k} - y_m\| < \varepsilon$.

In case, V is a finite dimensional normed linear space, each Cauchy sequence in S is convergent; however, the limit vector may not be in S. For our purpose, we then require S to be a closed subset of a finite dimensional normed linear space V so that each Cauchy sequence in S will have its limit vector in S.

With this little background from analysis, we essentially show that the fixed-point iteration with a contraction map defines a Cauchy sequence. But we rephrase it keeping its applications in mind.

Theorem 7.4 (Contraction mapping principle) *Let S be a nonempty closed subset of a finite dimensional normed linear space V. If $f : S \to S$ is a contraction, then f has a unique fixed point in S.*

Proof Denote the norm on V as $\| \cdot \|$. Let $f : S \to S$ be a contraction. Then, there exits a positive constant $c < 1$ such that

$$\|f(u) - f(v)\| < c \|u - v\| \quad \text{for all } u, v \in S.$$

Choose any vector $x_0 \in S$. Define the fixed-point iteration

$$x_0, \quad x_{n+1} = f(x_n) \quad \text{for } n \geq 0.$$

Since $f : S \to S$, each $x_j \in S$. Also, for any $m \geq 1$,

$$\|x_{m+1} - x_m\| = \|f(x_m) - f(x_{m-1})\| \leq c \|x_m - x_{m-1}\|.$$

Then,

$$\|x_{m+1} - x_m\| \leq c \|x_m - x_{m-1}\| \leq c^2 \|x_{m-1} - x_{m-2}\| \leq \cdots \leq c^m \|x_1 - x_0\|.$$

By the triangle inequality,

$$\begin{aligned} \|x_{m+k} - x_m\| &\leq \|(x_{m+k} - x_{m+k-1}) + \cdots + (x_{m+1} - x_m)\| \\ &\leq (c^{m+k-1} + \cdots + c^m)\|x_1 - x_0\| \end{aligned}$$

$$= c^m \frac{1 - c^k}{1 - c} \|x_1 - x_0\|$$

$$\leq \frac{c^m}{1 - c} \|x_1 - x_0\|. \tag{7.2}$$

As $0 < c < 1$ implies that $0 < c^k < 1$, we have $0 < 1 - c^k < 1$; thus, the last inequality holds.

Suppose $\varepsilon > 0$. Since $\lim\limits_{n \to \infty} \dfrac{c^m}{1 - c} = 0$, we have a natural number N such that for each $m > N$, $\dfrac{c^m}{1 - c} < \varepsilon$. That is,

$$\|x_{m+k} - x_m\| < \varepsilon \quad \text{for all } m > N, \ k > 0.$$

Therefore, $\{x_n\}_{n=0}^{\infty}$ is a Cauchy sequence in S. Since S is a closed subset of a finite dimensional normed linear space, this sequence converges, and the limit vector, say u, is in S.

Observe that since $\|f(x) - f(y)\| < c \|x - y\|$, the function f is continuous. Thus, taking limit of both the sides in the fixed-point iteration $x_{n+1} = f(x_n)$, we see that $u = f(u)$. Therefore, u is a fixed point of f.

For uniqueness of such a fixed point, suppose u and v are two fixed points of f. Then,

$$u = f(u), \quad v = f(v).$$

If $u \neq v$, then $0 < \|u - v\| = \|f(u) - f(v)\| < c \|u - v\| < \|u - v\|$ as $c < 1$. This is a contradiction. Therefore, $u = v$. ■

Contraction mapping principle can be used to solve linear or nonlinear equations as we have mentioned earlier. The trick is to write the equation in the form $x = f(x)$ so that such an f is a contraction on its domain. Next, we use the fixed-point iteration to solve the equation approximately.

It will be useful to have some idea on the possible error we may commit when we declare that a certain x_m in the fixed-point iteration is an approximation of the fixed point x. The two types of error estimates are as follows.

A priori error estimate: $\quad \|x_m - x\| \leq \dfrac{c^m}{1 - c} \|x_1 - x_0\|.$

A posteriori error estimate: $\quad \|x_m - x\| \leq \dfrac{c}{1 - c} \|x_m - x_{m-1}\|.$

The first one follows from the inequality $\|x_{m+k} - x_m\| \leq \dfrac{c^m}{(1 - c)} \|x_1 - x_0\|$ in (7.2), by taking $k \to \infty$. This error estimate is called a priori since without computing x_m explicitly, it gives information as to how much error we may commit by approximating x with x_m.

For the second one, observe that

$$\|x_m - x\| = \|f(x) - f(x_{m-1})\| \le c \|x - x_{m-1}\|. \tag{7.3}$$

Using the triangle inequality, we have

$$\|x - x_{m-1}\| \le \|x - x_m\| + \|x_m - x_{m-1}\| \le c \|x - x_{m-1}\| + \|x_m - x_{m-1}\|.$$

This implies

$$(1 - c) \|x - x_{m-1}\| \le \|x_m - x_{m-1}\|.$$

Using (7.3), we obtain

$$\|x - x_m\| \le c\|x - x_{m-1}\| \le \frac{c}{1 - c} \|x_m - x_{m-1}\|.$$

This error estimate is called a posteriori since it gives information about the error in approximating x with x_m only after x_m has been computed.

Example 7.7 A commonly used method to approximate a root of a nonlinear equation $f(x) = 0$ is the *Newton's method*, where the iteration is given by

$$x^0, \quad x^{n+1} = x^n - (f'(x^n))^{-1} f(x^n) \quad \text{for } n \ge 0.$$

We may consider $f(x)$ as a vector function, say, $f : \mathbb{R}^{n \times 1} \to \mathbb{R}^{n \times 1}$. That is,

$$f(x) = (f_1(x_1, \ldots, x_n), \ldots, f_n(x_1, \ldots, x_n)),$$

where $f_j : \mathbb{R}^{n \times 1} \to \mathbb{R}$ is a real valued function of n variables. Since the components of x are written as x_i with subscripts, we have written the mth approximation to x as x^m, using the superscript notation.

Newton's method can be posed as a fixed-point iteration with the iteration function g given by

$$g(x) = x - (f'(x))^{-1} f(x).$$

Here, $f'(x)$ is the Jacobian matrix

$$f'(x) = [a_{ij}] \in \mathbb{R}^{n \times n} \quad \text{with } a_{ij} = \frac{\partial f_i}{\partial x_j} \quad \text{for } i, j = 1, \ldots, n.$$

The resulting iteration $x^{n+1} = x^n - (f'(x^n))^{-1} f(x^n)$ is solved by setting $h^n = x^{n+1} - x^n$ which satisfies the linear system

$$f'(x^n)h^n = -f(x^n).$$

To prove that Newton's method converges to a solution, we should define S on which this g would be a contraction under suitable conditions on f.

Denote a root of the equation $f(x) = 0$ by u. Assume that f and its first and second partial derivatives are continuous, $f'(x)$ is invertible within a neighbourhood of radius $\delta > 0$ of the vector u and that in the same neighbourhood, $\|(f'(x))^{-1}\| \leq \alpha$ for some $\alpha > 0$. Using Taylor's formula and then taking the norm, we have

$$\|f(x) - f(u) - f'(x)(x - u)\| \leq \beta \|u - x\|^2 \quad \text{for some } \beta > 0.$$

As $f(u) = 0$, it implies that

$$\|g(x) - u\| = \|x - u - (f'(x))^{-1} f(x)\| \leq \beta \|(f'(x))^{-1}\| \, \|u - x\|^2.$$

We choose $\varepsilon < \min\{\delta, \ (\alpha\beta)^{-1}\}$ and define $S := \{x : \|u - x\| \leq \varepsilon\}$. Then,

$$\|g(x) - u\| \leq \alpha\beta \|u - x\|^2 \leq \alpha\beta\varepsilon \|u - x\| \leq \|u - x\|.$$

Therefore, $g : S \to S$. Also, if $\|u - x\| \leq \varepsilon$, then $c = \alpha\beta \|u - x\| < 1$. So, g is a contraction on S. We conclude that Newton's method starting with any x^0 within the ε neighbourhood of the root u will converge to u.

Observe that this feature of Newton's method requires a good initial guess x^0. If the initial guess is away from the desired root, then the method may not converge to the root. □

Exercises for Sect. 7.3

1. Determine the largest interval $S = \{x \in \mathbb{R} : a \leq x \leq b\}$ satisfying the property: for each $x_0 \in S$, the fixed-point iteration $x_{i+1} = x_i(2 - x_i)$ converges.
2. Show that the fixed-point iteration $x_{i+1} = x_i - x_i^3$ converges if the initial guess x_0 satisfies $-1 < x_0 < 1$.
3. Let $S = \{x \in \mathbb{R} : x \geq 0\}$. Define $f : S \to S$ by $f(x) = x + (1 + x)^{-1}$. Show that for all $x, y \in S$, $|f(x) - f(y)| < |x - y|$. Also, show that f is not a contraction.
4. Show that Newton's method for the function $f(x) = x^n - a$ for $x > 0$, $n > 1$, $a > 0$, converges to $a^{1/n}$.

7.4 Iterative Solution of Linear Systems

Let $A \in \mathbb{C}^{n \times n}$ be an invertible matrix. To solve the linear system $Ax = b$, we rewrite it as $x = x + C(b - Ax)$ for an invertible matrix C. We see that the solution of the linear system is a fixed point of $f(x)$, where

$$f(x) = x + C(b - Ax) \quad \text{for an invertible matrix } C.$$

By using this $f(x)$ as an iteration function, we may obtain an approximate solution for the linear system.

Notice that we could have rewritten $Ax = b$ as $x = x + (b - Ax)$. However, keeping an arbitrary invertible matrix C helps in applying the contraction mapping principle; see the following theorem.

Theorem 7.5 (Fixed-Point Iteration for Linear Systems) *Let* $\| \cdot \|$ *be a sub-multiplicative norm on matrices. Let* $A \in \mathbb{F}^{n \times n}$, $b \in \mathbb{F}^{n \times 1}$, *and let* $C \in \mathbb{F}^{n \times n}$ *be such that* $\| I - CA \| < 1$. *Then, the following iteration converges to the solution of the system* $Ax = b$:

$$x^0, \quad x^{m+1} = x^m + C(b - Ax^m) \quad for \ m \geq 0 \tag{7.4}$$

Proof First, we show that both A and C are invertible. On the contrary, if at least one of A or C is not invertible, then there exists a nonzero vector $x \in \mathbb{C}^{n \times 1}$ such that $CAx = 0$. Then,

$$\|x\| = \|x - CAx\| = \|(I - CA)x\| \leq \|I - CA\| \|x\| < \|x\|,$$

a contradiction. Hence, both A and C are invertible.

Consequently, there exists a unique solution of the linear system $Ax = b$.

Let $f : \mathbb{C}^{n \times 1} \to \mathbb{C}^{n \times 1}$ be defined by

$$f(x) = x + C(b - Ax) \quad \text{for } x \in \mathbb{C}^{n \times 1}.$$

If $x, y \in \mathbb{C}^{n \times 1}$, then

$$\begin{aligned} \| f(x) - f(y) \| &= \| x + C(b - Ax) - y - C(b - Ay) \| \\ &= \| (I - CA)(x - y) \| \leq \| I - CA \| \, \| x - y \|. \end{aligned}$$

Since $\| I - CA \| < 1$, f is a contraction on $\mathbb{C}^{n \times 1}$. By the contraction mapping principle, $f(x)$ has a unique fixed point in $\mathbb{C}^{n \times 1}$. If this fixed point is u, then $u = f(u)$ implies that $u = u + C(b - Au)$, or $C(b - Au) = 0$. Since C is invertible, we have $Au = b$. That is, the fixed point u is the solution of $Ax = b$.

Notice that the fixed-point iteration with the iteration function as $f(x)$ is nothing but Iteration (7.4). This completes the proof. ∎

Iteration (7.4) for solving a linear system is generically named as the *fixed-point iteration for linear systems*. However, it is not just a method; it is a scheme of methods. By fixing the matrix C and a sub-multiplicative norm, we obtain a corresponding method to approximate the solution of $Ax = b$.

For a given linear system, Iteration (7.4) requires choosing a suitable matrix C such that $\| I - CA \| < 1$ in some induced norm or the Frobenius norm. It may quite well happen that in one induced norm, $\| I - CA \|$ is greater than 1, but in another induced norm, the same is smaller than 1. For example, consider the matrix

$$B = \begin{bmatrix} 2/3 & 0 \\ 2/3 & 0 \end{bmatrix}.$$

Its row sum norm, column sum norm, the Frobenius norm, and the spectral norm tell different stories:

$$\|B\|_\infty = \tfrac{2}{3} + \tfrac{2}{3} = \tfrac{4}{3} > 1, \quad \|B\|_1 = \tfrac{2}{3} < 1, \quad \|B\|_F = \tfrac{\sqrt{8}}{3} < 1, \quad \|B\|_2 = \tfrac{\sqrt{8}}{3} < 1.$$

The last one is computed from the eigenvalues of B^*B, which are $\tfrac{8}{9}$ and 0.
  The spectral radius of B is (See Exercise 1.):

$$\rho(B) = \max\{|\lambda| : \lambda \text{ is an eigenvalue of } B\} = \tfrac{2}{3}.$$

That is, the norm of B in any induced norm is at least $\tfrac{2}{3}$.

Example 7.8 Consider the linear system $Ax = b$, where $A = [a_{ij}] \in \mathbb{C}^{n \times n}$ and $b \in \mathbb{C}^{n \times 1}$. Let $D = \text{diag}(a_{11}, \ldots, a_{nn})$, which consists of the diagonal portion of A as its diagonal and the rest of the entries are all 0. Suppose that no diagonal entry of A is 0, and that in some induced norm, or in the Frobenius norm, $\|I - D^{-1}A\| < 1$.
  For instance, suppose A is a **strict diagonally dominant** matrix, that is, the entries a_{ij} of A satisfy

$$|a_{ii}| > \sum_{j=1, j \neq i}^{n} |a_{ij}| \quad \text{for } i = 1, \ldots, n.$$

This says that in any row, the absolute value of the diagonal entry is greater than the sum of absolute values of all other entries in that row. Then,

$$\|I - D^{-1}A\|_\infty = \max\left\{ 1 - \sum_{j=1}^{n} |a_{ij}|/|a_{ii}| : 1 \leq i \leq n \right\}$$

$$= \max\left\{ \sum_{j=1, j \neq i}^{n} |a_{ij}|/|a_{ii}| : 1 \leq i, j \leq n \right\} < 1.$$

It follows that the iteration

$$x^0, \quad x^{m+1} = x^m + D^{-1}(b - Ax^m) \quad \text{for } m \geq 0$$

converges to the unique solution of $Ax = b$. This fixed-point iteration is called *Jacobi iteration*. □

  In Jacobi iteration, we essentially take D^{-1} as an approximation to A. In that case, we may view the lower triangular part of A to be a still better approximation to A. We try this heuristic in the next example.

Example 7.9 For $A = [a_{ij}] \in \mathbb{C}^{n \times n}$ and $b \in \mathbb{C}^{n \times 1}$, let $L = [\ell_{ij}]$, where

$$\ell_{ij} = a_{ij} \quad \text{for } i \geq j, \quad \text{and} \quad \ell_{ij} = 0 \quad \text{for } i < j.$$

Assume that A is invertible and that no diagonal entry of A is 0 so that L is also invertible. Suppose that in some induced norm or in Frobenius norm, $\|I - L^{-1}A\| < 1$. For instance, if A is strict diagonally dominant, then $\|I - L^{-1}A\|_\infty < 1$ as earlier. Then, the fixed-point iteration

$$x^0, \quad x^{m+1} = x^m + L^{-1}(b - Ax^m) \quad \text{for } m \geq 0$$

converges to the solution of $Ax = b$.

This fixed-point iteration is called *Gauss–Seidel iteration*. It can be shown that it converges for all positive definite matrices also.   □

Exercises for Sect. 7.4

1. For any matrix $A \in \mathbb{F}^{n \times n}$, the non-negative real number

$$\rho(A) = \max\{|\lambda| : \lambda \text{ is an eigenvalue of } A\}$$

   is called the *spectral radius* of A. Show that if $\|\cdot\|$ is any induced norm, then $\rho(A) \leq \|A\|$.
2. Solve the linear system $3x - 6y + 2z = 15$, $-4x + y - z = -2$, $x - 3y + 7z = 22$ by using

   (a) Jacobi iteration with initial guess as $(0, 0, 0)$.
   (b) Gauss–Seidel iteration with initial guess as $(2, 2, -1)$.

7.5 Condition Number

In applications, when we reach at a linear system $Ax = b$ for $A \in \mathbb{F}^{n \times n}$ and $b \in \mathbb{F}^{n \times 1}$, it is quite possible that there are errors in the matrix A and the vector b. The errors might have arisen due to modelling or due to computation of other parameters. Neglecting small parameters may lead to such errors. Or, you could have entered the data incorrectly due to various reasons such as the constants are irrational numbers but you need to feed in only floating point numbers, or by sheer mistake. It is also possible that the entries in A and b have been computed from numerical solution of another problem. In all such cases, we would like to determine the effect of small changes in the data (that is, in A and b) on the solution of the system and its accuracy.

In other words, we assume that we are actually solving a perturbed proble, and we would like to estimate the error committed due to perturbation. Even with negligible perturbation, the nature of the perturbed system can very much differ from the original one.

Example 7.10 Consider $A = \begin{bmatrix} 1 & 1 \\ 2 & 2 \end{bmatrix}$, $A' = \begin{bmatrix} 1+\delta & 1+\delta \\ 2 & 2 \end{bmatrix}$, $\tilde{A} = \begin{bmatrix} 1+\delta & 1 \\ 2 & 2 \end{bmatrix}$
and $b = \begin{bmatrix} 2 \\ 4 \end{bmatrix}$ for some $\delta \neq 0$.

With a small value of δ, say, $\delta = 10^{-10}$, the matrices A' and $\tilde{A}$ may be considered as small perturbations to the matrix A.

We find that the system $Ax = b$ has infinitely many solutions given by

$$\text{Sol}(A, b) = \left\{ \begin{bmatrix} \alpha \\ 2 - \alpha \end{bmatrix} : \alpha \in \mathbb{F} \right\}.$$

The system $A'x = b$ has no solutions. Whereas the system $\tilde{A}x = b$ has a unique solution $x_1 = -1/\delta$, $x_2 = 2 + 1/\delta$. $\square$

In analysing the effect of a perturbation on a solution, we require the so-called condition number of a matrix. Let $A \in \mathbb{F}^{n \times n}$ be an invertible matrix. Let $\| \cdot \|$ be an induced matrix norm or the Frobenius norm. Then, the **condition number** of A, denoted by $\kappa(A)$, is defined as

$$\kappa(A) = \|A^{-1}\| \, \|A\|.$$

If A is not invertible, then we define $\kappa(A) = \infty$.

Naturally, the condition number depends on the chosen matrix norm. We take the condition number of a non-invertible matrix as infinity due to the complete unpredictability of the nature of a solution of the corresponding linear system under perturbation.

Example 7.11 The inverse of $A = \begin{bmatrix} 1 & 1/2 & 1/4 \\ 1/2 & 1 & 1/4 \\ 1/4 & 1/2 & 1 \end{bmatrix}$ is $A^{-1} = \frac{2}{21} \begin{bmatrix} 14 & -6 & -2 \\ -7 & 15 & -2 \\ 0 & -6 & 12 \end{bmatrix}$.

If we use the maximum row sum norm, then

$$\|A\|_\infty = \tfrac{7}{4}, \quad \|A^{-1}\|_\infty = \tfrac{2}{21} \times 24 = \tfrac{16}{7}.$$

Thus, $\kappa(A) = \tfrac{7}{4} \times \tfrac{16}{7} = 4$.

If we use the maximum column sum norm, then

$$\|A\|_1 = \tfrac{7}{4}, \quad \|A^{-1}\|_1 = \tfrac{2}{21} \times 27 = \tfrac{18}{7}.$$

Consequently, $\kappa(A) = \tfrac{7}{4} \times \tfrac{18}{7} = \tfrac{9}{2}$. $\square$

Observe that the condition number of any matrix with respect to any induced norm is at least 1. For,

$$1 = \|I\| = \|A^{-1}A\| \le \|A^{-1}\| \, \|A\| = \kappa(A).$$

Therefore, we informally say that a matrix is **well-conditioned** if its condition number is close to 1; and it is **ill-conditioned** if its condition number is large compared to 1.

We wish to see how an error in the data leads to small or large errors in the solution of a linear system. The data in a linear system $Ax = b$ are the matrix A and the vector b. We consider these two cases separately.

First, suppose that A is invertible and there is an error in b. We estimate its effect on the solution, especially, on the relative error.

Theorem 7.6 *Let $A \in \mathbb{F}^{n \times n}$ be invertible. Let b, $b' \in \mathbb{F}^{n \times 1}$. Let x, $x' \in \mathbb{F}^{n \times 1}$ satisfy $x \ne 0$, $Ax = b$, and $Ax' = b'$. Then,*

$$\frac{\|x' - x\|}{\|x\|} \le \kappa(A) \frac{\|b' - b\|}{\|b\|}.$$

Proof $Ax = b$ and $Ax' = b'$ imply that $A(x' - x) = b' - b$. That is, $x' - x = A^{-1}(b' - b)$. Then,

$$\|x' - x\| \le \|A^{-1}\| \, \|b' - b\|.$$

Next, $Ax = b$ implies that $\|b\| \le \|A\| \, \|x\|$. Multiplying this with the previous inequality, we obtain

$$\|x' - x\| \, \|b\| \le \|A^{-1}\| \, \|b' - b\| \, \|A\| \, \|x\| = \kappa(A) \, \|b' - b\| \, \|x\|.$$

The required estimate follows from this.   ∎

Next, consider the case that b is fixed but A is perturbed. We estimate the relative error in the solution using the relative error in A.

Theorem 7.7 *Let $A \in \mathbb{F}^{n \times n}$ be invertible, $A' \in \mathbb{F}^{n \times n}$, and let $b \in \mathbb{F}^{n \times 1}$. Let x, $x' \in \mathbb{F}^{n \times 1}$ satisfy $Ax = b$, $x' \ne 0$, and $A'x' = b$. Then,*

$$\frac{\|x' - x\|}{\|x'\|} \le \kappa(A) \frac{\|A' - A\|}{\|A\|}.$$

Proof $Ax = b$ and $A'x' = b$ imply that $A(x' - x) + (A' - A)x' = 0$. It gives

$$x' - x = -A^{-1}(A' - A)x'.$$

Taking norms of both sides, we have

$$\|x' - x\| \le \|A^{-1}\| \, \|A' - A\| \, \|x'\| = \frac{\kappa(A)}{\|A\|} \, \|A' - A\| \, \|x'\|.$$

The required inequality follows from this.   ∎

The two estimates in Theorems 7.6–7.7 need to be interpreted correctly. Roughly, they say that when the condition number is large, the relative error will also be large, in general. They do not say that largeness of the condition number *ensures* large relative error.

In Theorem 7.6, the estimate is valid for any fixed b. It means that for some $b \in \mathbb{F}^{n \times 1}$ the relative error will be large when the condition number of A is large. Also, there may be some $b \in \mathbb{F}^{n \times 1}$ for which the relative error is small even if $\kappa(A)$ is large. Similar comment applies to Theorem 7.7.

When we compute the solution of a linear system, due to round-off and other similar factors, the computed solution may not be an exact solution. It means it may not satisfy the linear system. Suppose that x is the exact solution of the linear system $Ax = b$ and our computation has produced $\hat{x}$. Substituting in the equation, we find that $A\hat{x} \neq b$.

The vector $b - A\hat{x}$ is called the **residual** of the computed solution. The residual can always be computed a posteriori. The **relative residual** is the quantity $\dfrac{\|b - A\hat{x}\|}{\|x\|}$.

We wish to estimate the relative error in the computed solution using the relative residual. Again, the condition number comes of help.

Theorem 7.8 *Let $A \in \mathbb{F}^{m \times n}$ be invertible, $b \in \mathbb{F}^{n \times 1}$, and let $b \neq 0$. Let $x \in \mathbb{F}^{n \times 1}$ satisfy $x \neq 0$ and $Ax = b$. Let $\hat{x} \in \mathbb{F}^{n \times 1}$. Then,*

$$\frac{1}{\kappa(A)} \frac{\|b - A\hat{x}\|}{\|b\|} \leq \frac{\|x - \hat{x}\|}{\|x\|} \leq \kappa(A) \frac{\|b - A\hat{x}\|}{\|b\|}.$$

Proof $A(x - \hat{x}) = Ax - A\hat{x} = b - A\hat{x}$ implies $\|b - A\hat{x}\| \leq \|A\| \, \|x - \hat{x}\|$. Also, $x - \hat{x} = A^{-1}(b - A\hat{x})$. Therefore,

$$\frac{\|b - A\hat{x}\|}{\|A\|} \leq \|x - \hat{x}\| = \|A^{-1}(b - A\hat{x})\| \leq \|A^{-1}\| \, \|b - A\hat{x}\|.$$

Dividing by $\|x\|$, we have

$$\frac{\|b\|}{\|A\| \, \|x\|} \cdot \frac{\|b - A\hat{x}\|}{\|b\|} \leq \frac{\|x - \hat{x}\|}{\|x\|} \leq \frac{\|A^{-1}\| \, \|b\|}{\|x\|} \cdot \frac{\|b - A\hat{x}\|}{\|b\|}. \tag{7.5}$$

Since $Ax = b$, we get $\|b\| \leq \|A\| \, \|x\|$; and $x = A^{-1}b$ implies $\|x\| \leq \|A^{-1}\| \, \|b\|$. It follows that

$$\frac{\|A^{-1}\| \, \|b\|}{\|x\|} \leq \|A^{-1}\| \, \|A\|, \quad \text{and} \quad \frac{1}{\|A^{-1}\| \, \|A\|} \leq \frac{\|b\|}{\|A\| \, \|x\|}. \tag{7.6}$$

Then, from (7.5)–(7.6), we obtain

$$\frac{1}{\|A^{-1}\| \, \|A\|} \cdot \frac{\|b - A\hat{x}\|}{\|b\|} \leq \frac{\|x - \hat{x}\|}{\|x\|} \leq \|A^{-1}\| \, \|A\| \frac{\|b - A\hat{x}\|}{\|b\|}.$$

Now, the required estimate follows.   ■

Observe that (7.5) can be used to obtain a lower bound as well as an upper bound
on the relative residual.

The estimate in Theorem 7.8 says that the relative error in the approximate solution
of $Ax = b$ can be as small as $\dfrac{\|b\|}{(\|A\|\,\|x\|)}$ times the relative residual, or it can be as

large as $\dfrac{\|A^{-1}\|\,\|b\|}{\|x\|}$ times its relative residual. It also says that when the condition
number of A is close to 1, the relative error and the relative residual are close to each
other; while the larger is the condition number of A, the relative residual provides
less information about the relative error.

Example 7.12 The inverse of $A = \begin{bmatrix} 1.01 & 0.99 \\ 0.99 & 1.01 \end{bmatrix}$ is $A^{-1} = \begin{bmatrix} 25.25 & -24.75 \\ -24.75 & 25.25 \end{bmatrix}$.
Using the maximum row sum norm, we see that

$$\|A\|_\infty = 2, \quad \|A^{-1}\|_\infty = 50, \quad \kappa(A) = 100.$$

The linear system $Ax = [2,\ 2]^t$ has the exact solution $x = [1,\ 1]^t$. We take an
approximate solution as $\hat{x} = [1.01\ 1.01]^t$. The relative error and the relative residual
are given by

$$\frac{\|x - \hat{x}\|_\infty}{\|x\|_\infty} = \frac{\|[0.01,\ 0.01]^t\|_\infty}{\|[1,\ 1]^t\|_\infty} = 0.01,$$

$$\frac{\|[2,\ 2]^t - A\hat{x}\|_\infty}{\|x\|_\infty} = \frac{\|[-0.02,\ -0.02]^t\|_\infty}{\|[1,\ 1]^t\|_\infty} = 0.02.$$

It shows that even if the condition number is large, the relative error and the
relative residual can be of the same order.

We change the right hand side to get the linear system $Ax = [2,\ -2]^t$. It has
the exact solution $x = [100,\ -100]^t$. We consider an approximate solution $\hat{x} =
[101,\ -99]^t$. The relative error and the relative residual are

$$\frac{\|x - \hat{x}\|_\infty}{\|x\|_\infty} = \frac{\|[-1,\ -1]^t\|_\infty}{\|[100,\ -100]^t\|_\infty} = 0.01,$$

$$\frac{\|[2,\ -2]^t - A\hat{x}\|_\infty}{\|x\|_\infty} = \frac{\|[-100.01,\ -100.01]^t\|_\infty}{\|[100,\ -100]^t\|_\infty} = 1.0001.$$

That is, the relative residual is 100 times the relative error.   □

In general, when $\kappa(A)$ is large, the system $Ax = b$ is ill-conditioned for at least one
choice of b. There might still be choices for b so that the system is well-conditioned.

Example 7.13 Let $A = \begin{bmatrix} 1 & 10^5 \\ 0 & 1 \end{bmatrix}$. Its inverse is $A^{-1} = \begin{bmatrix} 1 & -10^5 \\ 0 & 1 \end{bmatrix}$. Using the absolute row sum norm, we have $\kappa(A) = \|A\|_\infty \|A^{-1}\|_\infty = (10^5 + 1)^2$.

The system $Ax = [1, \ 1]^t$ has the solution $x = [1 - 10^5, \ 1]^t$.

Changing the vector b to $b' = [1.001, \ 1.001]^t$, the system $Ax = b'$ has the solution $x' = [1.001 - 10^5 - 10^2, \ 1.001]^t$. We find that the relative change in the solution and the relative residual are as follows:

$$\frac{\|x - x'\|_\infty}{\|x\|_\infty} = \frac{\|0.001[1 - 10^5, \ 1]^t\|_\infty}{\|[1 - 10^5, \ 1]^t\|_\infty} = 0.001,$$

$$\frac{\|b - b'\|_\infty}{\|b\|_\infty} = \frac{\|[0.001 \ 0.001]^t\|_\infty}{\|[1 \ 1]^t\|_\infty} = 0.001.$$

It shows that even if the condition number is large, the system can be well-conditioned. $\qquad\square$

Exercises for Sect. 7.5

1. Using the spectral norm, show that an orthogonal matrix has the condition number 1.
2. Using the norm $\| \cdot \|_2$ for matrices, compute the condition number of the matrix $\begin{bmatrix} 3 & -5 \\ 6 & 1 \end{bmatrix}$.
3. Let $A, B \in \mathbb{F}^{n \times n}$. Show that $\kappa(AB) \le \kappa(A)\kappa(B)$.
4. Using the norm $\| \cdot \|_2$ for $n \times n$ matrices, show that $\kappa(A) \le \kappa(A^*A)$.

7.6 Matrix Exponential

The exponential of a complex number is usually defined by a series, which is an infinite sum. Since addition is defined only for two numbers, and hence by induction, for a finite number of them, the infinite sum requires a new definition. Instead of telling that an infinite sum such as a series has so and so value, we say that it converges to such and such a number. Similarly, we may define when an infinite sum of vectors or matrices converges.

Let $\{A_i\}_{n=1}^\infty$ be a sequence of $m \times n$ matrices with complex entries. We say that the series $\sum_{i=1}^\infty A_i$ **converges to** an $m \times n$ matrix A iff for each $\varepsilon > 0$, there exists a natural number n_0 such that if $k > n_0$, then $\|\sum_{i=1}^k A_i - A\| < \varepsilon$.

Notice that for a square matrix A, any sub-multiplicative matrix norm satisfies the property that $\|A^k\| \le \|A\|^k$ for all natural numbers k. Thus, when each A_j is an $n \times n$ matrix and $\| \cdot \|$ is sub-multiplicative, it can be shown (by using Cauchy sequences) that the series of matrices $\sum_{j=1}^\infty A_j$ converges to some matrix if the series of non-negative real numbers $\sum_{j=1}^\infty \|A_j\|$ converges to some real number. Using this, the exponential of a square matrix A, which we denote by e^A, is defined as follows:

$$e^A = \sum_{i=0}^{\infty} \frac{A^i}{i!} = I + A + \frac{1}{2!}A^2 + \frac{1}{3!}A^3 + \cdots$$

The computation of e^A is simple when A is a diagonal matrix. For,

$$D = \text{diag}(d_1, \ldots, d_n) \quad \text{implies} \quad e^D = \text{diag}(e^{d_1}, \ldots, e^{d_n}).$$

If A is similar to a diagonal matrix, then we may write $A = PDP^{-1}$ for some invertible matrix P. Then, $A^i = P\,D^i\,P^{-1}$; consequently, $e^A = P\,e^D\,P^{-1}$. This way, exponential of any diagonalizable matrix can be computed.

In general, we may use the Jordan form of A for obtaining a closed-form expression for e^A. Suppose that $A = PJP^{-1}$, where J is a matrix in Jordan form. Here,

$$J = \text{diag}(J_1, \ldots, J_k),$$

where J_i are the Jordan blocks in the form

$$J_i = \begin{bmatrix} \lambda & 1 & & & \\ & \lambda & 1 & & \\ & & \ddots & \ddots & \\ & & & & 1 \\ & & & & \lambda \end{bmatrix}$$

for an eigenvalue λ of A. Then,

$$e^A = Pe^J P^{-1} = P\,\text{diag}(e^{J_1}, \ldots, e^{J_k})\,P^{-1}.$$

If J_i in the above form is of order ℓ, then e^{J_i} is in the following form:

$$e^{J_i} = e^{\lambda} \begin{bmatrix} 1 & 1 & 1/2! & 1/3! & \cdots & & 1/(\ell-1)! \\ & 1 & 1 & 1/2! & \cdots & & 1/(\ell-2)! \\ & & 1 & 1 & \cdots & & 1/(\ell-3)! \\ & & & \ddots & & & \vdots \\ & & & & 1 & 1 & 1/2! \\ & & & & & 1 & 1 \\ & & & & & & 1 \end{bmatrix}$$

The exponential of a matrix comes often in the context of solving ordinary differential equations. Consider the initial value problem

$$\frac{dx}{dt} = Ax, \quad x(t_0) = x^0$$

where $x(t) = [x_1(t), \cdots, x_n(t)]^t$, $A \in \mathbb{R}^{n \times n}$, and $x^0 = [a_1, \cdots, a_n]^t \in \mathbb{R}^{n \times 1}$. The solution of this initial value problem is given by

$$x(t) = e^{(t-t_0)A} x^0 = e^{tA} e^{-t_0 A} x^0.$$

The matrix e^{tA} may be computed via the Jordan form of A as outlined above.

There is an alternative. We wish to find out n linearly independent vectors $v_1, \ldots, v_n$ such that the series $e^{tA} v_i$ can be summed exactly. If this can be done, then taking these as columns of a matrix B, we see that

$$B = e^{tA} \begin{bmatrix} v_1 & \cdots & v_n \end{bmatrix}.$$

The matrix $\begin{bmatrix} v_1 & \cdots & v_n \end{bmatrix}$ is invertible since its columns are linearly independent. Then, using its inverse, we can compute e^{tA}. We now discuss how to execute this plan.

For any scalar λ, $t(A - \lambda I)(t\lambda I) = (t\lambda I)t(A - \lambda I)$. Thus, we have

$$e^{tA} = e^{t(A-\lambda I)} e^{t\lambda I} = e^{t(A-\lambda I)} e^{\lambda t} I = e^{\lambda t} e^{t(A-\lambda I)}.$$

Hence, if $v \in \mathbb{R}^{n \times 1}$ is such that $(A - \lambda I)^m v = 0$, then

$$e^{tA} v = e^{\lambda t} e^{t(A-\lambda I)} v = e^{\lambda t} \left[v + t(A - \lambda I)v + \cdots + \frac{t^{m-1}}{(m-1)!} (A - \lambda I)^{m-1} v \right].$$

That is, for a generalized eigenvector v, the infinite sum turns out to be a finite sum. The question is that can we choose n such generalized eigenvectors for A? From the discussions about Jordan form, we know that the answer is affirmative. In fact, the following is true:

Let λ be an eigenvalue of A with algebraic multiplicity m. If the linear system $(A - \lambda I)^k x = 0$ has $r < m$ number of linearly independent solutions, then $(A - \lambda I)^{k+1}$ has at least $r + 1$ number of linearly independent solutions.

Our plan is to obtain m number of linearly independent generalized eigenvectors associated with an eigenvalue λ of algebraic multiplicity m. Then, we take together all such generalized eigenvectors for obtaining the n linearly independent vectors v so that $e^{tA} v$ can be summed exactly. We proceed as follows.

1. Find all eigenvalues of A along with their algebraic multiplicity. For each eigenvalue λ, follow Steps 2–6:
2. Determine linearly independent vectors v satisfying $(A - \lambda I)v = 0$.
3. If m number of such vectors are found, then write them as $v_1, \ldots, v_m$, and set $w_1 := e^{\lambda_1 t} v_1, \ldots, w_m := e^{\lambda_n t} v_m$. Go to Step 7.
4. In Step 2, suppose that only $k < m$ number of vectors v_i could be obtained. To find additional vectors, determine all vectors v such that $(A - \lambda I)^2 v = 0$ but $(A - \lambda I)v \neq 0$. For each such vector v, set the corresponding w as

$$w := e^{\lambda t}\big(v + t(A - \lambda I)v\big).$$

5. If n number of vectors w_j could not be found in Steps 3 and 4, then determine all vectors v such that $(A - \lambda I)^3 v = 0$ but $(A - \lambda I)^2 v \neq 0$. Corresponding to each such v, set

$$w := e^{\lambda t}\Big(v + t(A - \lambda I)v + \frac{t^2}{2!}(A - \lambda I)^2 v\Big).$$

6. Continue to obtain more vectors w considering vectors v that satisfy $(A - \lambda I)^{j+1} v = 0$ but $(A - \lambda I)^j v \neq 0$, and by setting

$$w := e^{\lambda t}\Big(v + t(A - \lambda I)v + \frac{t^2}{2!}(A - \lambda I)^2 v + \cdots + \frac{t^j}{j!}(A - \lambda I)^j v\Big).$$

7. Assume that for each eigenvalue λ of algebraic multiplicity m, we have obtained $v_1, \ldots, v_m$ and $w_1, \ldots, w_m$. Together, we have obtained n number of linearly independent v_is and n number of linearly independent w_is. Then, set

$$B = \begin{bmatrix} w_1 & \cdots & w_n \end{bmatrix}, \quad C = \begin{bmatrix} v_1 & \cdots & v_n \end{bmatrix}, \quad e^{tA} = BC^{-1}.$$

Further, the vectors $w_1, \ldots, w_n$ are n linearly independent solutions of the system of differential equations $dx/dt = Ax$.

Example 7.14 Let $A = \begin{bmatrix} 1 & 0 & 0 \\ 1 & 1 & 0 \\ 0 & 0 & 3 \end{bmatrix}$. Its eigenvalues, with multiplicities, are 1, 1, 3.

For $\lambda = 1$, we determine all nonzero vectors v such that $(A - \lambda I)v = 0$. If $v = [a, \ b, \ c,]^t$, then

$$(A - 1\,I)v = \begin{bmatrix} 0 & 0 & 0 \\ 1 & 0 & 0 \\ 0 & 0 & 2 \end{bmatrix}\begin{bmatrix} a \\ b \\ c \end{bmatrix} = \begin{bmatrix} 0 \\ a \\ 2c \end{bmatrix} = \begin{bmatrix} 0 \\ 0 \\ 0 \end{bmatrix}.$$

It implies that $a = 0, \ c = 0$, and b is arbitrary. We choose

$$v_1 = [0, \ 1, \ 0]^t, \quad w_1 = e^t[0, \ 1, \ 0]^t.$$

Notice that the eigenvalue 1 has algebraic multiplicity 2 but we got only one linearly independent eigenvector. To compute an additional generalized eigenvector for this eigenvalue, we find vectors v such that $(A - 1\,I)^2 v = 0$ but $(A - 1\,I)v \neq 0$. With $v = [a, \ b, \ c]^t$, it gives

$$(A - 1\,I)^2 v = \begin{bmatrix} 0 & 0 & 0 \\ 1 & 0 & 0 \\ 0 & 0 & 2 \end{bmatrix}^2 \begin{bmatrix} a \\ b \\ c \end{bmatrix} = \begin{bmatrix} 0 & 0 & 0 \\ 1 & 0 & 0 \\ 0 & 0 & 2 \end{bmatrix}\begin{bmatrix} 0 \\ a \\ 2c \end{bmatrix} = \begin{bmatrix} 0 \\ 0 \\ 2c \end{bmatrix} = \begin{bmatrix} 0 \\ 0 \\ 0 \end{bmatrix}.$$

It implies that a, b are arbitrary and $c = 0$. Moreover, it should also satisfy $(A - \lambda I)v \neq 0$. Taking any vector linearly independent with v_1 will be such a choice. Thus, we choose $a = 1$, $b = 0$, $c = 0$. That is,

$$v_2 - [1, \ 0, \ 0]^t, \quad w_2 = e^t(v_2 + t(A - 1\,I)v_2) = [e^t, \ te^t, \ 0]^t.$$

For $\lambda = 3$, $(A - \lambda I)[a, \ b, \ c]^t = 0$ implies that $-2a = 0 = a - 2b$. That is, $a = 0$, $b = 0$, and c is arbitrary. Thus, we choose

$$v_3 = [0, \ 0, \ 1]^t, \quad w_3 = e^{3t}[0, \ 0, \ 1]^t.$$

We have got three linearly independent vectors v_1, v_2, v_3 and the corresponding w_1, w_2, w_3. We put these vectors as columns of matrices B and C to obtain

$$B = \begin{bmatrix} w_1 & w_2 & w_3 \end{bmatrix} = \begin{bmatrix} 0 & e^t & 0 \\ e^t & te^t & 0 \\ 0 & 0 & e^{3t} \end{bmatrix}, \quad C = \begin{bmatrix} v_1 & v_2 & v_3 \end{bmatrix} = \begin{bmatrix} 0 & 1 & 0 \\ 1 & 0 & 0 \\ 0 & 0 & 1 \end{bmatrix}.$$

Then,

$$e^{tA} = BC^{-1} = \begin{bmatrix} 0 & e^t & 0 \\ e^t & te^t & 0 \\ 0 & 0 & e^{3t} \end{bmatrix}\begin{bmatrix} 0 & 1 & 0 \\ 1 & 0 & 0 \\ 0 & 0 & 1 \end{bmatrix} = \begin{bmatrix} e^t & 0 & 0 \\ te^t & e^t & 0 \\ 0 & 0 & e^{3t} \end{bmatrix}, \quad e^A - \begin{bmatrix} e & 0 & 0 \\ e & e & 0 \\ 0 & 0 & e^3 \end{bmatrix}.$$

Further, we observe that w_1, w_2 and w_3 are three linearly independent solutions of the system of differential equations $dx/dt = Ax$. □

Exercises for Sect. 7.6

1. Let λ be an eigenvalue of a matrix A. Show that e^λ is an eigenvalue of the exponential matrix e^A.
2. Let $D = \text{diag}(\lambda_1, \ldots, \lambda_n)$. Show that $e^D = \text{diag}(e^{\lambda_1}, \ldots, e^{\lambda_n})$.
3. Let $A = \begin{bmatrix} \lambda & 1 & 0 \\ 0 & \lambda & 1 \\ 0 & 0 & \lambda \end{bmatrix}$. Show that $e^{tA} = e^{\lambda t}\begin{bmatrix} 1 & t & t^2/2 \\ 0 & 1 & t \\ 0 & 0 & 1 \end{bmatrix}$.
4. Let A be a Jordan block of order n with diagonal entries as λ. Let $B = A - \lambda I$. Show that $B^n = 0$. Further, prove that

$$e^{tA} = e^{\lambda t}\left(I + tB + \frac{t^2}{2!}B^2 + \cdots + \frac{t^{n-1}}{(n-1)!}B^{n-1}\right).$$

5. Compute e^{tA} in the following cases, where A is given by

(a) $\begin{bmatrix} 0 & 1 \\ -1 & 0 \end{bmatrix}$ (b) $\begin{bmatrix} 1 & 1 & 0 & 0 \\ 0 & 1 & 0 & 0 \\ 0 & 0 & 1 & 0 \\ 0 & 0 & 0 & 1 \end{bmatrix}$ (c) $\begin{bmatrix} 1 & 1 & 0 & 0 \\ 0 & 1 & 1 & 0 \\ 0 & 0 & 1 & 0 \\ 0 & 0 & 0 & 1 \end{bmatrix}$ (d) $\begin{bmatrix} 1 & 1 & 0 & 0 \\ 0 & 1 & 1 & 0 \\ 0 & 0 & 1 & 1 \\ 0 & 0 & 0 & 1 \end{bmatrix}$

6. Compute e^A, where A is given by

(a) $\begin{bmatrix} 1 & 1 \\ -1 & -1 \end{bmatrix}$ (b) $\begin{bmatrix} 3 & 4 \\ -2 & -3 \end{bmatrix}$ (c) $\begin{bmatrix} 1 & 1 & 1 \\ -1 & -1 & -1 \\ 1 & 1 & 1 \end{bmatrix}$

7. Let $A = \begin{bmatrix} 0 & 0 \\ 1 & 0 \end{bmatrix}$, $B = \begin{bmatrix} 1 & 0 \\ 1 & 0 \end{bmatrix}$. Show that $\text{tr}(e^{A+B}) \neq \text{tr}(e^A e^B)$.

8. Find e^A, where $A \in \mathbb{R}^{5 \times 5}$ with each entry as 1.

9. Determine e^{tA} if $A^2 = 7A$.

7.7 Estimating Eigenvalues

As you have seen, manual computation of eigenvalues of an arbitrary matrix using its characteristic polynomial is impossible. Many numerical methods have been devised to compute required number of eigenvalues approximately. This issue is addressed in numerical linear algebra. Prior to these computations, it is often helpful to have some information about the size or order of magnitude of the eigenvalues in terms of the norms of some related vectors or matrices.

Theorem 7.9 *Let $\| \cdot \|$ be an induced norm on $\mathbb{F}^{n \times n}$, and let λ be an eigenvalue of a matrix $A \in \mathbb{F}^{n \times n}$. Then, $|\lambda| \leq \|A\|$.*

Proof Let v be an eigenvector associated with the eigenvalue λ of the matrix A. Since $Av = \lambda v$, we have $\|Av\| = |\lambda|\,\|v\|$. Then,

$$|\lambda| = \frac{\|Av\|}{\|v\|} \leq \text{lub}\left\{ \frac{\|Av\|}{\|v\|} : v \in \mathbb{F}^{n \times 1},\ v \neq 0 \right\} = \|A\|. \qquad \blacksquare$$

In particular, since the row sum norm and the column sum norm are induced norms, we have the following computationally simple upper bounds on the eigenvalues of a matrix:

$$|\lambda| \leq \max_i \left(\sum_{j=1}^{n} |a_{ij}| \right), \quad |\lambda| \leq \max_j \left(\sum_{i=1}^{n} |a_{ij}| \right),$$

where λ is any eigenvalue of the matrix $A = [a_{ij}] \in \mathbb{F}^{n \times n}$.

In the complex plane, an inequality such as $|\lambda| < r$ gives rise to a disc of radius r centred at 0, in which λ lies. Thus, Theorem 7.9 says that all eigenvalues of A lie inside the disc of radius $\|A\|$ centred at the origin. To improve this result, we introduce some notation and some terminology.

Let $A = [a_{ij}] \in \mathbb{F}^{n \times n}$. Let $r_i(A)$ denote the sum of absolute values of all entries in the ith row except the diagonal entry. That is,

$$r_i(A) = |a_{i1}| + \cdots + |a_{i(i-1)}| + |a_{i(i+1)}| + \cdots + |a_{in}| = \sum_{j=1,\, j \neq i}^{n} |a_{ij}|.$$

The ith **Geršgorin disc** $D_i(A)$ of A is defined as

$$D_i(A) = \{z \in \mathbb{C} : |z - a_{ii}| \le r_i(A)\}.$$

Thus, there are n number of Geršgorin discs; one for each row of A.

Theorem 7.10 (Geršgorin Discs) *All eigenvalues of a matrix lie inside the union of its Geršgorin discs.*

Proof Let λ be an eigenvalue of a matrix $A = [a_{ij}] \in \mathbb{F}^{n \times n}$ with an associated eigenvector v. Then, $v \ne 0$ but $(A - \lambda I)v = 0$. The vector v has n components. Suppose the ith component has the largest absolute value. That is, if $v = [b_1, \cdots, b_n]^t$, then $|b_i| \ge |b_j|$ for each $j \ne i$. Write the equality $(A - \lambda I)v = 0$ in detail, and consider the ith equality in it. It looks like

$$a_{i1}b_1 + \cdots + a_{i(i-1)}b_{i-1} + (a_{ii} - \lambda)b_i + a_{i(i+1)}b_{i+1} + \cdots + a_{in}b_n = 0.$$

Bringing the ith term to one side and taking absolute values, we have

$$|a_{ii} - \lambda| \, |b_i| \le \sum_{j=1, \, j \ne i}^{n} |a_{ij}b_j| = \sum_{j=1, \, j \ne i}^{n} |a_{ij}| \, |b_j| \le |b_i| \sum_{j=1, \, j \ne i}^{n} |a_{ij}| = |b_i| \, r_i(A).$$

Then, $|\lambda - a_{ii}| \le r_i(A)$. That is, $\lambda \in D_i(A)$. We see that corresponding to each eigenvalue λ there exists a row i of A such that $\lambda \in D_i$. Therefore, each eigenvalue of A lies in $D_1(A) \cup \cdots \cup D_n(A)$. ∎

Recall that a matrix $A = [a_{ij}] \in \mathbb{F}^{n \times n}$ is called strict diagonally dominant iff $|a_{ii}| > r_i(A)$ for each $i = 1, \ldots, n$. Look at the proof of Theorem 7.10. If A is not invertible, then 0 is an eigenvalue of A. With $\lambda = 0$, we obtain the inequality $|a_{ii}| \le r_i(A)$ for some row index i. Thus, *each strict diagonally dominant matrix is invertible.*

Example 7.15 Consider the matrix $A = \begin{bmatrix} 0 & 3 & 2 & 3 & 3 \\ -1 & 7 & 2 & 1 & 1 \\ 2 & 1 & 0 & 1 & 1 \\ 0 & -1 & 1 & 0 & 1 \\ 1 & -1 & 2 & 1 & 0 \end{bmatrix}$.

The Geršgorin discs are specified by complex numbers z satisfying

$$|z| \le 11, \quad |z - 7| \le 5, \quad |z| \le 5, \quad |z| \le 3, \quad |z| \le 5.$$

The first disc contains all others except the second. Therefore, all eigenvalues lie inside the union of discs $|z| \le 11$ and $|z - 7| \le 5$.

Notice that A^t and A have the same eigenvalues. It amounts to taking the Geršgorin discs corresponding to the columns of A. Here, they are specified as follows:

$$|z| \leq 4, \quad |z - 7| \leq 6, \quad |z| \leq 7, \quad |z| \leq 6, \quad |z| \leq 6.$$

As earlier, it follows that all eigenvalues of A lie inside the union of the discs $|z| \leq 7$ and $|z - 7| \leq 6$.

Therefore, all eigenvalues of A lie inside the intersection of the two regions obtained earlier as unions of Geršgorin discs. It turns out that this intersection is the union of the discs $|z| \leq 7$ and $|z - 7| \leq 5$. □

Further sharpening of Geršgorin's theorem is possible. One such useful result is the following:

Let $A \in \mathbb{F}^{n \times n}$. Let $k \in \{1, \ldots, n\}$. If k of the Geršgorin discs for A are disjoint from the other $n - k$ Geršgorin discs, then exactly k of the eigenvalues of A lie inside the union of these k discs.

There are many improvements on Geršgorin's theorem giving various kinds of estimates on eigenvalues. You may see Varga [15].

Exercises for Sect. 7.7

1. Using Geršgorin discs, determine the regions where all eigenvalues of the following matrices lie: (a) $\begin{bmatrix} 5 & 2 & 4 \\ -2 & 0 & 2 \\ 2 & 4 & 7 \end{bmatrix}$ (b) $\begin{bmatrix} 0 & -10 & 1 \\ 5 & 2 & 0 \\ -8 & 10 & 12 \end{bmatrix}$

   (c) $\begin{bmatrix} i & 1-i & 0 \\ i/2 & 0 & 2i \\ 1+i & 2+i & 3+4i \end{bmatrix}$

2. Give an example of a 2×2 matrix, where at least one of its eigenvalues lies on the boundary of the union of Geršgorin discs.

7.8 Problems

1. Let c and s be real numbers with $c^2 + s^2 = 1$. Show that $\|A\|_F = \sqrt{n}$, where

$$A = \begin{bmatrix} -1 & c & c & \cdots & c & c \\ 0 & -s & cs & \cdots & cs & cs \\ 0 & 0 & -s^2 & \cdots & cs^2 & cs^2 \\ & & & \vdots & & \\ 0 & 0 & 0 & \cdots & -s^{n-2} & cs^{n-2} \\ 0 & 0 & 0 & \cdots & 0 & -s^{n-2} \end{bmatrix} \in \mathbb{F}^{n \times n}.$$

2. Let $A \in \mathbb{F}^{m \times n}$. Let $U \in \mathbb{F}^{m \times m}$ be unitary. Show that $\|A\|_F = \|UA\|_F$.

3. Let $A \in \mathbb{F}^{m \times n}$ have rank r. Use the previous problem and the SVD of A to deduce that $\|A\|_F^2 = \sum_{i=1}^{r} s_i^2$, where s_i is the ith positive singular value of A.

4. Show that $\|A + B\|_F^2 = \|A\|_F^2 + 2\operatorname{tr}(A^t B) + \|B\|_F^2$ for $n \times n$ matrices A and B with real entries.

5. Let A be an invertible matrix. Show that $\|A^{-1}\|_2$ is the reciprocal of the smallest positive singular value of A.

6. Let $A \in \mathbb{F}^{n\times n}$ be a hermitian matrix, and let $v \in \mathbb{F}^{n\times 1}$ be a nonzero vector. Show that the Rayleigh quotient $\rho_A(v) \in \mathbb{R}$.

7. Let A be a hermitian $n \times n$ matrix with eigenvalues $\lambda_1 \geq \cdots \geq \lambda_n$, and let v be any column vector of size n. Show that the Rayleigh quotient $\rho_A(v)$ satisfies $\lambda_1 \geq \rho_A(v) \geq \lambda_n$.

8. Let $A = \begin{bmatrix} 5 & 4 & -4 \\ 4 & 5 & 4 \\ -4 & 4 & 5 \end{bmatrix}$. Let $v = \begin{bmatrix} 1 \\ t \\ 1 \end{bmatrix}$ where $t \in \mathbb{R}$.

   (a) The Rayleigh quotient $\rho_A(v)$ depends on t; so call it $f(t)$. Compute $f(t)$.
   (b) Compute the minimum and the maximum values of $f(t)$ to estimate the largest and the smallest eigenvalues of A.
   (c) Does A have an eigenvector v whose first and third components are same?

9. Show that if the fixed-point iteration $x_{i+1} = x_i^2 - 2$ converges to 2, then there exists a natural number n_0 such that for each $n > n_0$, $x_n = 2$.

10. Consider finding a root of the equation $4x^2 - e^x = 0$ by using the fixed-point iteration $x_{i+1} = \frac{1}{2}e^{x_i/2}$. Show the following:

    (a) If $0 \leq x_0 \leq 1$, then the iteration converges to the root in that lies between 0 and 1.
    (b) The iteration does not converge to the root that lies between 4 and 5 with any x_0.

11. Let $p(t)$ be a polynomial with real coefficients. Assume that $p(t)$ has only real zeros. Let z be the largest zero of $p(t)$. Show that Newton's method with an initial guess $x_0 > z$ converges to z.

12. In which matrix norm, the condition number is the ratio of its largest singular value to its smallest positive singular value?

13. Let $A, B \in \mathbb{F}^{n\times n}$ such that $AB = BA$. Prove that $e^{A+B} = e^A e^B$.

14. Give examples of 2×2 matrices A and B such that $e^{A+B} \neq e^A e^B$.

15. Let λ be an eigenvalue of A with an associated eigenvector x. Show that x is also an eigenvector of e^A. What is the corresponding eigenvalue?

16. Show that e^A is invertible for any $A \in \mathbb{F}^{n\times n}$.

17. Let A be a real symmetric positive definite matrix. Show that e^A is also such a matrix.

18. Let $A \in \mathbb{R}^{n\times n}$. Is e^A always symmetric? Is e^A always positive definite?

19. Let $A \in \mathbb{R}^{n\times n}$. Show that each column of e^{At} is a solution of the differential equation $dx/dt = Ax$.

20. Let $A \in \mathbb{R}^{n\times n}$. Let $f_i(t)$ be the solution of the initial value problem $dx/dt = Ax$, $x(0) = e_i$, the ith standard basis vector of $\mathbb{R}^{n\times 1}$. Show that $e^{At} = [f_1(t), \ldots, f_n(t)]$.

References

1. M. Braun, *Differential Equations and Their Applications*, 4th edn. (Springer, New York, 1993)
2. R.A. Brualdi, The Jordan canonical form: an old proof. Am. Math. Mon. **94**(3), 257–267 (1987)
3. S.D. Conte, C. de Boor, *Elementary Numerical Analysis: An algorithmic approach* (McGraw-Hill Book Company, Int. Student Ed., 1981)
4. J.W. Demmel, *Numerical Linear Algebra* (SIAM Pub, Philadelphia, 1996)
5. F.R. Gantmacher, *Matrix Theory*, vol. 1–2 (American Math. Soc., 2000)
6. G.H. Golub, C.F. Van Loan, *Matrix Computations*, Hindustan Book Agency, Texts and Readings in Math. - 43, New Delhi (2007)
7. R. Horn, C. Johnson, *Matrix Analysis* (Cambridge University Press, New York, 1985)
8. P. Lancaster, M. Tismenetsky, *The Theory of Matrices*, 2nd edn. (Elsevier, 1985)
9. A.J. Laub, *Matrix Analysis for Scientists and Engineers* (SIAM, Philadelphia, 2004)
10. S.J. Leon, *Linear Algebra with Applications*, 9th edn. (Pearson, 2014)
11. D. Lewis, *Matrix Theory* (World Scientific, 1991)
12. C. Meyer, *Matrix Analysis and Applied Linear Algebra* (SIAM, Philadelphia, 2000)
13. R. Piziak, P.L. Odell, *Matrix Theory: From Generalized Inverses to Jordan Form* (Chapman and Hall/CRC, 2007)
14. G. Strang, *Linear Algebra and Its Applications*, 4th edn. (Cengage Learning, 2006)
15. R.S. Varga, *Geršgorin and His Circles*, Springer Series in Computational Mathematics, vol. 36 (Springer, 2004)

© The Editor(s) (if applicable) and The Author(s), under exclusive license to Springer
Nature Switzerland AG 2021
A. Singh, *Introduction to Matrix Theory*,
https://doi.org/10.1007/978-3-030-80481-7

Index

Printed in the United States
by Baker & Taylor Publisher Services